Microplastic Pollution: Causes, Effects and Control

Edited By

Rahul Singh

School of Bioengineering and Biosciences
Lovely Professional University
Punjab, India

&

Neeta Raj Sharma

School of Bioengineering and Biosciences
Lovely Professional University
Punjab, India

Microplastic Pollution: Causes, Effects and Control

Editors: Rahul Singh and Neeta Raj Sharma

ISBN (Online): 978-981-5165-10-4

ISBN (Print): 978-981-5165-11-1

ISBN (Paperback): 978-981-5165-12-8

Published by Bentham Science Publishers Pte. Ltd. Singapore. All Rights Reserved.

First published in 2023.

need for a court order if at any point you breach any terms of this License Agreement. In no event will any delay or failure by Bentham Science Publishers in enforcing your compliance with this License Agreement constitute a waiver of any of its rights.

3. You acknowledge that you have read this License Agreement, and agree to be bound by its terms and conditions. To the extent that any other terms and conditions presented on any website of Bentham Science Publishers conflict with, or are inconsistent with, the terms and conditions set out in this License Agreement, you acknowledge that the terms and conditions set out in this License Agreement shall prevail.

Bentham Science Publishers Pte. Ltd.
80 Robinson Road #02-00
Singapore 068898
Singapore
Email: subscriptions@benthamscience.net

**BENTHAM
SCIENCE**

CONTENTS

FOREWORD

Since the time plastics became an inseparable part of our daily life, we have produced unbeatable 8.4 billion tons of plastics. Rivers are contaminated with more than 2.4 million tons of synthetic polymers including microplastics. The term ''Microplastic'' was first coined in 2004 by Richard Thompson, a marine biologist. Plastic particles within the size range of 5mm − 1 μm are considered microplastics, whereas particles less than 1 μm have been termed as nano-plastics. Polyethylene, polypropylene, polyethylene terephthalate and polyamides comprise the majority of polymer types. These plastic particles can act as carriers of other toxic environmental contaminants such as, polychlorinated biphenyls (PCBs), polycyclic aromatic hydrocarbons (PAHs), plasticizers, pesticides, heavy metals, pharmaceuticals, and poly-fluoroalkyl substances, like perfluorooctanoic acid (PFOA) and perfluorooctane sulfonic acid (PFOS). We are not talking about their presence in just rivers and oceans; these plastic particles are present in much of the air that we breath, the water that we drink, and the food that we eat. In a recent report, the U.S. Geological Survey titled bluntly, "It is raining plastic", researchers describe how they found plastics in 90 percent of the rainwater samples. In this special issue on, 'Microplastic Pollution, Causes, Effects and Control, Professor Neeta Raj Sharma, and associates', describe the source of microparticles (MP), analytical techniques for the determination of MPs, and the impact of MPs on flora, fauna, aquatic and soil environments, with an example of their possible ill effects on human health. The publication of this comprehensive monograph is timely and will be well received by the scientific community.

Gundu H. R. Rao
Institute of Engineering Medicine, and Lillehei Heart Institute
University of Minnesota
Minneapolis, Minnesota
USA

PREFACE

In today's time, microplastic is known as an emerging pollutant. It was first recognized in an ocean, but as the search and analysis continued, it was claimed to be present in fresh water, groundwater, soil and even in the air. Microplastics in the environment can be of two types, the primary source which is made intensively in cosmetic products and the secondary source which is formed due to the degradation of plastics. As the search is increasing, the amount of degraded plastic and the negative impact seem to be increasing.

There is a lot of confusion about the name of microplastics, because its size starts from < 5 mm. Like its name, its sampling and analysis are also quite complex. As far as the impact is concerned, it is not only harmful in itself, but it also takes the form of a multiple-stressor by accumulating many hydrophobic contaminants in the water.

The editors would first like to thank all contributing to this important topic. We, thank you to Nafiaah Naqash who contributed significantly to the compilation of this book. We also thank the publication team wholeheartedly, from whose side we got full support, and the result is in front of you in the form of this book.

Rahul Singh
School of Bioengineering and Biosciences
Lovely Professional University
Punjab, India

&

Neeta Raj Sharma
School of Bioengineering and Biosciences
Lovely Professional University
Punjab, India

List of Contributors

Arun Karnwal	Department of Microbiology, School of Bioengineering and Biosciences, Lovely Professional University, Phagwara-144411, Punjab, India
Dhriti Kapoor	Department of Botany, School of Bioengineering and Biosciences, Lovely Professional University, Phagwara (Punjab), India
Dhriti Sharma	Department of Botany, School of Bioengineering and Biosciences, Lovely Professional University, Phagwara (Punjab), India
Francesco Fazio	Department of Veterinary Science, University of Messina, Messina, Italy
Gautam Priyadarshi	Department of Earth & Environmental Science, KSKV Kachchh University, Bhuj-370001, Gujarat, India
Harpreet Kaur	Department of Chemistry, Lovely Professional University (LPU), Phagwara, 144411 Punjab, India
Imtiaz Ahmed	Fish Nutrition Research Laboratory, Department of Zoology, University of Kashmir, Srinagar-190006, Jammu and Kashmir, India
Inderpal Devgon	Department of Microbiology, School of Bioengineering and Biosciences, Lovely Professional University, Phagwara-144411, Punjab, India
Ishtiyaq Ahmad	Fish Nutrition Research Laboratory, Department of Zoology, University of Kashmir, Srinagar-190006, Jammu and Kashmir, India
Khushboo	Department of Microbiology, School of Bioengineering and Biosciences, Lovely Professional University, Phagwara-144411, Punjab, India
Kuljit Kaur	Department of Chemistry, Lovely Professional University (LPU), Phagwara, 144411 Punjab, India
Manpreet Kaur Somal	Department of Biotechnology, School of Bioengineering and Biosciences, Lovely Professional University, Phagwara-144411, Punjab, India
Monita Dhiman	Khalsa College for Women, Civil Lines, Ludhiana, Punjab, India
Mrugesh Trivedi	Department of Earth & Environmental Science, KSKV Kachchh University, Bhuj-370001, Gujarat, India
Mukesh Kumar	Department of Microbiology, School of Bioengineering and Biosciences, Lovely Professional University, Phagwara-144411, Punjab, India
Nafiaah Naqash	School of Bioengineering and Biosciences, Lovely Professional University, Phagwara-144411, Punjab, India
Perumal Muthukumar	Department of Geology, V.O. Chidambaram College, Tuticorin-628008, Tamilnadu, India
Quseen Mushtaq Reshi	Fish Nutrition Research Laboratory, Department of Zoology, University of Kashmir, Srinagar-190006, Jammu and Kashmir, India
Rahul Singh	School of Bioengineering and Biosciences, Lovely Professional University, Phagwara-144411, Punjab, India
Reetika Rani	School of Bioengineering and Biosciences, Lovely Professional University, Phagwara-144411, Punjab, India

Rohan Samir Kumar Sachan	Department of Microbiology, School of Bioengineering and Biosciences, Lovely Professional University, Phagwara-144411, Punjab, India
Ritu Bala	Department of Microbiology, School of Bioengineering and Biosciences, Lovely Professional University, Phagwara-144411, Punjab, India
Sekar Selvam	Department of Geology, V.O. Chidambaram College, Tuticorin-628008, Tamilnadu, India
Sadguru Prakash	Department of Zoology, M.L.K. (P.G.) College, Balrampur, UP, India
Sagar Prajapati	Department of Earth & Environmental Science, KSKV Kachchh University, Bhuj-370001, Gujarat, India
Sarabjeet Kaur	Mehr Chand Mahajan DAV College for Women, Sector- 36, Chandigarh, India
Savita Bhardwaj	Department of Botany, School of Bioengineering and Biosciences, Lovely Professional University, Phagwara (Punjab), India
Suraya Partap Singh	Department of Zoology, Government Degree College Kathua, Kathua, Jammu and Kashmir, India
Tunisha Verma	Department of Botany, School of Bioengineering and Biosciences, Lovely Professional University, Phagwara (Punjab), India

CHAPTER 1

Tools and Techniques to Analyse Microplastic Pollution in Aquatic and Terrestrial Ecosystems

Gautam Priyadarshi[1], **Sagar Prajapati**[1] and **Mrugesh Trivedi**[1,*]

[1] *Department of Earth & Environmental Science, KSKV Kachchh University, Bhuj-370001, Gujarat, India*

Abstract: The estimation of microplastic pollution in the terrestrial and aquatic ecosystem is carried out by quantification and identification of the contaminated environment. Microplastic estimation consists of various steps such as sampling, visualization and quantification. Generally, the planktonic net, bongo net, manta net, and neuston net have been used for water sampling. While, grab samplers, tweezers, tablespoons, trowels, shovels, spatulas, or hand picking methods have been used for soil and sediment sampling. The biological sample from the study sites comprises the direct collection of the whole organism or its colony as a sampling unit. However all samples are required to be processed further to extract the microplastic using techniques such as filtration, density extraction, digestion, and magnetic & electrostatic extraction. The digestion method is used for direct characterization such as thermal gravimetric analysis. The identification of microplastic is based on microscopic images which provide the shape, size, colour, and texture of the microplastic surface. Visual identification using microscopes is time-consuming and susceptible to human error as well as a risk of misidentification, which leads to underestimation or overestimation of microplastic pollution. Spectroscopic methods such as ATR-FTIR, μ-FTIR and Raman spectroscopy provide identification and quantification of synthetic polymer. Advance combined analytical techniques have been reported during the last few years such as portable micro-Raman, SEM-FTIR, Pyr-GC-MS, TGA-DSC, and PEE. Priority and care are essential concerning the sampling, storage and handling microplastic samples for the QA/QC for accurate analysis. The present chapter aims to provide a comprehensive overview of the current knowledge of tools and techniques used for microplastic inquiries from an environmental sample.

Keywords: Aquatic ecosystem, Biota, Density extraction, Environment, Extraction, Electrostatic separation, Filtration, FTIR, Grab sampler, Identification, Magnetic extraction, Microplastic, Planktonic Net, Pyr-GC-MS, QA/QC, Quantification, Raman spectroscopy, SEM, Sampling, Sediment, Sample processing, Terrestrial ecosystem.

---

* **Corresponding author Mrugesh Trivedi:** Department of Earth & Environmental Science, KSKV Kachchh University, Bhuj-370001, Gujarat, India; E-mail: drmrugesh.trivedi@gmail.com

Rahul Singh and Neeta Raj Sharma (Eds.)

INTRODUCTION

Plastic pollutants with a fragment size of less than 5 mm are called Microplastics (MPs). They are caused by the physical or chemical degradation of several types of plastics in both aquatic and terrestrial ecosystems [1, 2, 3]. The first report on MPs in the marine ecosystem was found in 1972 with two different polymers including dine rubber and polystyrene. After this event, to date, MPs have been reported throughout the world (192 countries) however, only 44 countries have researched related to MPs. They have been found from various environmental samples including terrestrial and marine-based processes, including domestic and industrial drainage [4], maritime activities agricultural runoff [5] and wastewater treatment plants (WWTPs) effluent [6, 7]. Many recent investigations have revealed that MP pollution has come into our food, sea salt [8] and potable water as well [5, 9, 10]. Increasing MP pollution poses a risk to humans and the environment. The global oceans have also been littered with microplastic pollution as shown in Fig. (1). There are various levels of MPs's environmental exposure through ingestion of food, breathing air and dermal contact with textile and dust particles from the air. This may cause health issues [11]. Due to the MPs exposure to the environment, all the ecosystems may cause toxicity including oxidative stress, inflammatory lesions and increased risk of neoplasia *i.e.* abnormal growth of cells or tissues [12]. Therefore, the assessment of MP based on its types and abundance in each sphere of the environment is necessary. In this context, the challenge is to undertake a scientific investigation to minimize ambiguity in risk assessments of known or expected impacts of MPs to develop adequate strategies for its control. There are various analytical techniques including Imaging (to determine the shape, size colour, and texture) and spectroscopy (to identify chemical components/composition) that have been used to estimate the characterization of MPs [13]. Various studies have been published regarding the tools and techniques to analyse microplastic pollution from terrestrial and aquatic ecosystems [6, 14]. The various aspects of MPs analysis such as sample collection, handling, storage, processing, extraction, characterization and quantification of MPs from various components of the environment are discussed in the chapter. To avoid misidentification and reduce or avoid the cross-contamination of MPs during the various stages of analysis, quality assurance and control (QA/QC) is also mentioned along with some recommendations.

Global distribution of microplastics

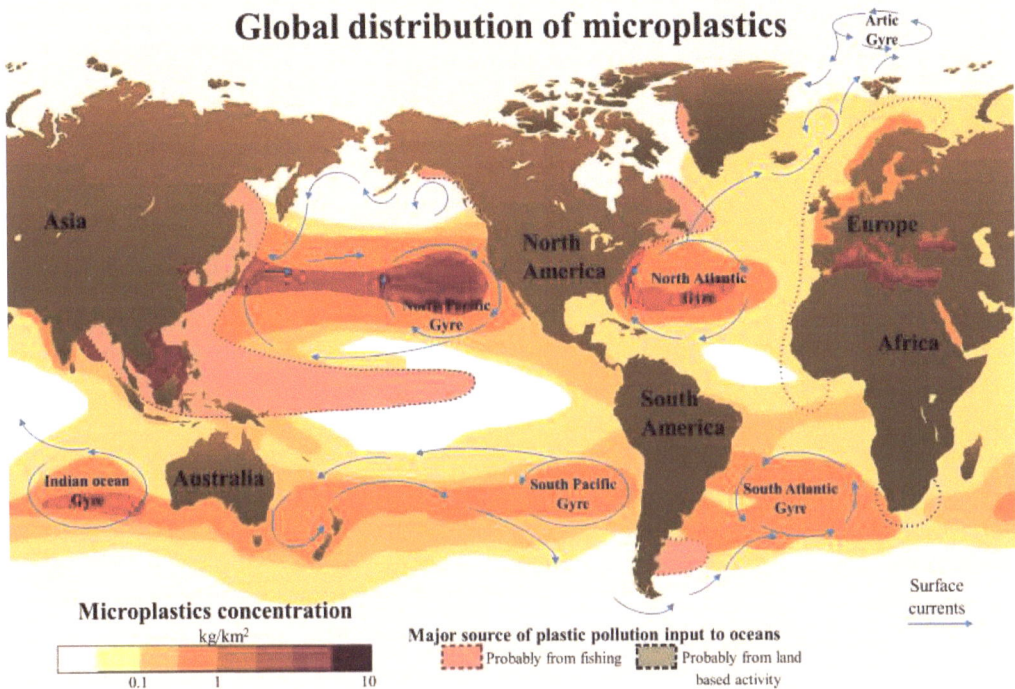

Fig. (1). Global distribution of microplastics in oceans. (Redrawn from "Marine Litter Vital Graphics").

SAMPLING OF MPs

There are mainly three approaches for sample collection including selective samples, bulk and volume-reduction. Selective sampling in which, items visible to the naked eye are directly extracted from the environment, such as on the surface of the water or sediment. This approach is simpler, however, there is a drawback that it is more obvious and sometimes heterogeneous types of particles and fibres are ignored because they are in a mixture with beach debris and other living organisms [15]. It is usually used at beaches and lakeshores [16]. Bulk sampling is the method in which the entire sample is taken without reducing its volume. When the sample size is too large, it is not possible to reduce the volume of the sample or identification of MPs is not possible through the naked eye, it is a more suitable technique. The majority of the biological samples are collected using this method [17, 18]. Sometimes, this sampling method is not representing the entire population but provides only a portion or a bunch of the population. To overcome this problem of sampling representation, volume reduction sampling method is being used to cover a large area. It reduces the volume of the entire population (bulk sample) until only the specific items of interest for further analysis remains.

It is a highly used method for water sampling [19]. The water, sediment and biological organism or its organ or tissue are generally collected as a sample to conduct MPs analysis.

Water Sample

MPs can directly (primary) or by the fragmentation of plastic particles (secondary), enters into any water body from the ocean to surface water (stream, lake, river, pond, and artificial pond) and groundwater as well. The water is a dynamic system where the MPs could be dispersed or transported with the turbulence and flow of the currents based on their size, shape and density [20]. Some climatic and physiochemical factors such as temperature, density and seasonal variation along the location of sample collection may affect the quality and quantity of the MPs sample. The riverine and marine water sampling is different as per above said physiochemical and climatic parameters. Water sampling can be done in two manners. One is surface water (planktonic sampling) and another one is water column (vertical sampling from top to bottom). Most of the planktonic sampling is being conducted using the bongo net, planktonic net, manta net and neuston net [21, 22]. It is a volume reduction method, where one can filter a large volume of water to collect the specific size of the sample. Generally, nets with a mesh width of 50-3000 μm were used [15]. Most of the nets are expensive, but it is widely used as it provides a huge volume to filter. To collect a lower quantity of water, a quick and lower expensive method is pumping, Niskin bottles, jars, buckets, Stainless-steel sieves and a Rotating Drum Sampler are used [21]. The result of MPs in water is usually given in mg/ L^{-1} or in per unit per L.

Sediment Sample

Sediments have been considered the long-term sink for MPs. The majority of sediment samples are collected from sandy beaches and deep oceans in marine ecosystems. The specific tidal zone sampled on a beach varied considerably among studies; some covered the entire extent of the beach, from the intertidal to the supralittoral zone. Bulk sampling is an excellent method to collect smaller MPs. However, bulk samples usually contain large amounts of unwanted substances, which increase the workload. Various sampling tools are being used for instance tweezers, tablespoons, trowels, shovels, spatulas or hand picking [23]. Sample collection from the seafloor requires a grab sampler. National Oceanic and Atmospheric Administration of USA (NOAA) recommends the use of 400 g per replicate, followed by drying and weighing to adjust the results, while some researchers have used the top 5 cm of the sediment surface [12]. The results of

MPs in sediments are reported in the number of microplastics per unit dry or wet weight (g or kg) area (m^2) or Volume (mL, L or m^3) from the sediments [24].

Biota

The evidence regarding the MPs in marine organisms and humans as well is continuously increasing. To analyse the MPs from the living organisms, individual specimens are sampled from various habitats and at different trophic levels. The phytoplankton and zooplankton have been reported using the above-mentioned vertical net tows from a depth of 250, or 10 m off the seafloor bottom, using Bongo nets. Fish species can be obtained with the pelagic net, trawl, electro fisher, or from the local fishermen, crustaceans and bivalves such as mussels and oysters can be acquired by hand, with a mussel trawl, or directly purchased from a store. The currently widely used qualification units of microplastics in aquatic organisms include the number of microplastics by weight and the number of microplastics per individual or the percentage of individuals containing ingested microplastics.

EXTRACTION OF MPS AND SAMPLE PREPARATION

The collected sample must be passed through the process of purification or prepared for quantitative analysis. It includes extraction, filtration, separation and digestion [25].

Flotation (Extraction by Density)

The collected water samples have to be first screened using a steel-wire sieve to eliminate debris. The collected particulates are resuspended in milli-Q water. Density separation methods use solutions with a wide range of densities, including but not limited to sodium chloride (NaCl, 1.2 g/cm^3), zinc chloride (ZnCl$_2$, 1.5–1.7 g/cm^3), and sodium iodide (NaI, 1.6–1.8 g/cm^3). The solution is directly added to the extracted MPs. As the MPs have a lower density, they will float on the surface. The sample is filtered using a glass microfilter. The filter paper is generally oven-dried at a moderate temperature (50-55°C). It is stored in a glass petri dish till further analysis [18]. For sediment extraction, a similar saturated solution is being used. 1:4 W/V solution and freeze-dried sediments are interacted to extract MPs from the sediments. To improve the extracted product, multiple replicates are needed. 30% hydrogen peroxide (H_2O_2) is also used to oxidise the organic matter. The entire sample is allowed to stay for one day after shaking. The supernatant (MPs containing) is separated using a filter paper.

Sieving or Filtration

Various sizes of sieves have been also reported to eliminate the unwanted compounds from the samples. The use of sieves with different mesh sizes allows for distinguishing size categories of microplastics. Sometimes, small-sized MPs can pass through larger mesh sizes of sieves, which leads to an underestimation of the MPs abundance. To avoid this problem, various types of filter paper are being used such as Glass-fibre, polycarbonate membrane, paper, nitrocellulose, and silicon filters are used for FT-IR spectroscopic analysis. The plastic particles are separated from the supernatant obtained by density separation by passing the solution that contains the plastic particles over a filter, usually aided by a vacuum. More usually, a filter paper with 1 to 1.6 μm pore size is also used [26, 27, 28].

Electrostatic Separation

Another separation technique is the electrostatic separation of MPs; a novel approach to separate the MPs from the solid material or dry sediment samples. The electrostatic separators divide the sample into non-conductive and conductive particles. As common plastics are non-conductive, they are separated from other metal particles. The dry sample should be used for the proper separation of plastic and non-plastic materials [29]. A vibrating conveyor that rotates a finely grounded metal drum that is transported to the area of a highly charged (35 kV) electrode is used for separation. The separated MPs trays are being emptied continuously, therefore the unlimited sample quantity can be run for separation. Some of the reviews have explained this method from a simple device to the recycling of industrial units. The advantage of this method is to provide almost 90- 100% of recovery from the soil or sediment sample in the range of 65 to 5 μm [30]. However, it does not apply to organic materials because it produces aggregates, which will reduce the capacity of the separation processes [31]. The application of this method to extra small particles must be verified, because the adhesive forces to the metal drum and scraper may be higher than the gravity force, which may be leading to huge losses of the small MPs to the final sample [32]. However, it has been a reliable, less time-consuming, cost-effective method that is easy to use for the separation of MPs from the large size of mineral-rich environmental samples [33, 34].

Magnetic Extraction

Separating MPs from bulk samples is difficult, however, plastics can be extracted by a magnet using the hydrophobic surface of MPs. Fe nanoparticles were used to extract MPs in 92% from various environments [34, 35]. The fine MPs were successfully extracted from dilute suspensions using this method [36]. Authors [35] have experimented on different polymers of MPs on collected sediments, which showed the recovery rates are from 49 to 90 per cent for medium-sized and up to 90 per cent for the smaller sized MPs. A recent investigation showed the removal and separation of MPs from the environmental samples using magnetic materials such as magnetic biochar [37] and magnetic carbon nanotubes [38].

Digestion

The collected sample containing organic matter may interrupt the analysis of MPs. The majority of biological samples are made up of organic material. These organic molecules are removed using the digestion method. Various acid and alkali solutions are being used to digest the organic matter based on the required polymer identification and estimation analysis [39]. For acid digestion, HNO_3 (nitric acid) is generally used. It provides the highest digestion (94-98%) of biological compounds compared to the other oxidizing agents such as HCl (Hydrochloric acid), NaOH (sodium hydroxide) and H_2O_2 [40]. Alkali digestion utilizes KOH (potassium hydroxide) and NaOH solution to digest the biological material. Increasing the concentration of the alkali solution with heating can increase the rate of digestion [41]. However, in chemical digestion, increasing the concentration can degrade the polymer such as Polyethylene terephthalate (PET), Polycarbonate, cellulose acetate and polyvinyl chloride [42]. To overcome the problem of MPs digestion, 15% H_2O_2 is used, which will not affect the Polyethylene and Polystyrene. Application of 30- 35% H_2O_2 requires a week to digest organic matter. The enzymatic degradation of MPs was carried out [41, 43] using cellulase, chitinase, lipase, and protease in a small sample size. It is a time-consuming process that requires proper incubation conditions and is limited to the small sample size. Microwave-assisted digestion was used [44, 45] to digest the biological sample to extract MPs. 100 mg of freeze-dried sample was digested in the microwave using closed Teflon vessels with HNO_3 (65%) for 45 minutes. 2 M NaOH solution was to neutralize the extract before filtration. Microwave digestion required less time but lower recovery (only 34%) as compared to the enzymatic digestion. Ultrasonication is also used with the combined digestion method which is more efficient in the fish intestine. It does not distort the polymer but may cause decoloration [46].

CHARACTERIZATION AND QUANTIFICATION

Visual Identification

Larger size MPs (1- 5 mm), such as plastic pellets, and plastic manufacturing waste, have been identified since the earliest studies during ocean surface water sampling or in beach sediments by visual identification with the naked eye based on the plastic colour and shape [47]. To visually identify the MPs [30], three rules are suggested which are as follows:

i. *No cellular or organic structures are visible,*

ii. *Fibres should be equally thick throughout their entire length variation,*

iii. *Particles should exhibit clear and homogeneous colors throughout variation.*

However, one may find some variation based on the morphology of MPs and other debris. Large MPs can be identified by visual inspection [48] while, the smaller MPs are generally sorted out under a compound microscope and Dissecting microscope using slides, coverslips, gridded filter paper and glass petri dish [49]. The identified MPs particles are sorted in a sorting chamber also known as the Bogorov counting chamber.

Manual Counting (Optical Microscope)

Manual Counting under the optical microscope is a widely used quantification method that identifies MPs and classifies categories based on their morphology (size, colour and shape). The large MPs are generally isolated and identified at the same time in Petri plates with forceps using the visual method (naked eyes), while the small MPs also can be identified, due to colourful plastic fragments and pre-production resin pellets [50]. Further identification of MPs can be done by Stereo-zoom (dissecting) microscopy, especially for the size range of hundreds of microns. It provides magnified images with more detailed surface, texture, and structural information of MPs. It will be more important for identifying confusing plastic-like particles. This method involves some limitations such as being labour intensive and having a chance to misidentify and count errors. The study [51] found that SEM analysis of MPs showed 20% of the particles of aluminium silicate from coal ash, which were initially considered as MPs by visual observation. The spectroscopic analytical technique can increase the accuracy of manual counting for the identification and quantification of MPs.

Polymer Identification

There are various techniques (microscopy and spectroscopy) that have been used for the detailed characterization of MPs. The high magnification and high-resolution images can be taken using Scanning Electron Microscope (SEM). It helps to determine the microscopic (nano) size, shape, texture and structure. It is used to analyse the weathering progress of MPs in natural conditions by their surface morphology [16]. The elemental composition (mainly inorganic) of MPs can be identified using energy-dispersive X-ray Spectroscopy (EDS) [52]. The elemental composition is used to identify the possible source of MPs. The specific chemical structural and functional group attached to any MPs can be analysed by Fourier transform infrared (FTIR) spectroscopy. It has been broadly used to identify the MPs based on the surface morphology however, it does not apply to a large number of samples as it requires time and sample preparation [53]. A significantly ($p < 0.05$) greater number of fragments were detected by FT-IR than by microscope. Two possible explanations exist for the underestimated abundance of fragments using a microscope. First, many transparent or white fragments were identified as synthetic polymers, such as polyethylene (PE) and polypropylene (PP) by FT-IR, but were not counted as microplastics using the microscope [54]. Apart from this, the focal plane array (FPA) FTIR microscopy (μ-FTIR) provides imaging data to analyse the MPs [24]. The analysis cost is dependent on the size of the MPs, as the size decreases, the sophisticated equipment analysis charges are increased. Thus, the first thing is to consider the size of MPs to select appropriate analysis techniques. For the analysis of MPs from the size of 100 microns to 5 mm, the FTIR- Attenuated Total Reflection (ATR) is required. The ATR works by pressing the MPs sample in front of a transparent crystal (diamond) and the infrared (IR) light passes through the sample. MPs sample absorbs some energy and reflects IR to the crystal which provides an IR spectrum of the polymer. The spectra are a further reference to the IR library database to examine the specific spectra of various polymers such as polyethylene, 273 polypropylene, polystyrene, or polyvinyl chloride of MPs [49, 55]. There is a wide range (5 mm to >20 μ) of FTIR combination techniques such as FTIR-ATR, FTIR Small spot ATR, and μ-FTIR with an Imaging microscope available for thorough identification of any plastic polymer (ThermoFisher, 2018).

Raman Spectroscopy

Another spectroscopic technique is Raman spectroscopy that is working on monochromatic light passing on the sample and measures the changes in photons on the sample surface. It provides a non-destructive characterization with structural information of the polymer that is present in the MPs sample. The MPs analysis of polymer was reported [56] to analyse MPs. It is a surface technique therefore, large MPs can be analysed and also coupled with a microscope. μ-Raman spectroscopy allows the analysis of lower range (<1 μ) MPs [57]. Furthermore, it can be combined with a confocal laser scanning microscope which enables to analyse MPs from biological tissues [41]. The disadvantage of this technique is the staining colour and pigment or associated chemicals may interfere with the identification and estimation of MPs [58]. Now portable Raman spectroscopy is available to capture the images from the isolated sample without a fluorescent dye [59]. It provides a cost-effective and rapid estimation of MPs from the aquatic ecosystem. It can detect the MPs from 100 nm in size and 40 to 0.15 μg/mL concentration [27, 60].

Emerging Techniques (Pyr GC-MS, GC-MS, TGA, DSC and PEE)

Some recent studies have reported pyrolysis gas chromatography coupled with mass spectrometry (Pyr-GC-MS) was used to identify the polymer from MPs in terrestrial and aquatic ecosystems. Pyrolysis is a combustion-based technique, where a sample is burned in an anoxia condition. The combusted material will undergo gas chromatography (GC) for the separation of chemical compounds, and it will be further identified using mass spectrometry (MS) detectors. This kind of combined analysis technique is attached to different detection spectroscopy such as Differential Scanning Calorimetry (DCS) GC-MS and Pyr-GC-MS [21, 52, 54]. It is a qualitative method that determines only the mass of the polymer but will not give any quantitative analysis.

Another analytical technique Thermo Gravimetric Analysis (TGA) uses the same principle and identifies the stability of any compound at various ranges of higher temperature (20 to 800 °C). As the plastic polymers show different thermal stability, they exhibited different physicochemical properties and their weight is lost [61]. The thermal decomposition of the polymer present in MPs gives a fingerprint identification based on temperature variation. It is more suitable for MPs that have low solubility, and those that are not easily dissolved and extracted. DSC is another thermal analysis technique that measures the heat flux, melting enthalpy and crystallization kinetics of MPs polymer at various increasing temperatures. It requires a reference sample to measure the change in the heat flux and provides an identification of primary MPs such as polyethene (PE) and polypropylene (PP) polymers with known characteristics [62]. The drawback of this technique is that it is unable to identify the mixture of MPs polymer with a similar or adjacent melting point. Therefore, the reproducibility of MPs polymer is influenced by the experimental conditions such as the amount of sample, size, and shape of the sample, heating rate and type of atmosphere where the sample is heated [63]. TGA determines the loss of MPs polymer weight concerning the temperature increase, whereas the degrading MPs polymer continuously changes its enthalpy. It is unable to measure the change in enthalpy however, it can be measured by DSC. Therefore, the combined technique (TGA-DSC, TGA-FTIR, TGA-MS) has been used [61, 64, 65] to identify the polypropylene and polyethylene polymers. These types of combined TGA-based analytic techniques are playing a crucial role in the fast identification of MPs polymers.

A novel technique is also reported using Pressurised Fried Extraction (PPE). It uses various solvents such as methanol, methanol, hexane and dichloromethane to extract volatile and semi-volatile plastic compounds under sub-critical temperature and pressure [66, 67]. It is also used to extract water-insoluble or slightly water-soluble organic compounds from soils, sediments and wastewater samples. It is considered to be one of the most accurate and reliable techniques as it utilizes the entire volume of samples for the quantitative extract. However, it is a sample destructive method that is unable to give morphological characteristics.

Universally accepted MPs analytical methods have not been developed, more precisely a Quality Assurance and Quality Control (QA/QC). The identification and quantification of MPs from aquatic and terrestrial ecosystems involve the chance of errors. It leads to misidentification, either underestimation or overestimation of MPs. The standard protocol may reduce the error margin and false identification. Each stage of analysis (sampling, storage, extraction & sample processing, identification and quantification) needs to be validated using blanks and standards. The method blank needs to be tested to estimate the potential contamination of laboratory reagent and working standards [68, 69, 70]. Very few studies have shown validation and control samples to maintain the quality of research [45, 67, 71]. Some of the literature reviews [23, 50, 72, 73] have already identified the gap of QA/QC and show some of the recommendations and precautions during the collection and handling of the MPs samples such as using the glass and metal equipment during sampling, storage and handling of MPs samples. The analytical techniques used such as (SEM, Raman, FTIR, TGA, and GC-MS) should have known detection limits and a reliable database source for MPs identification and quantification [74, 75]. The use of synthetic cloths should be avoided (aprons and cleaning cloths). All the laboratory equipment and surface should be cleaned with a tissue paper and alcohol. All the working reagents and chemical solutions must be filtered to avoid cross-contamination. The working area and laboratory should be air-conditioned as even air may contain MPs fibres [2, 76]. Covering the sample and applying fume hood can significantly reduce the chance of contamination as reported in a study [77].

CONCLUSION

Analysis of MPs from terrestrial and aquatic ecosystems involves a 4-step analysis process including sampling, extraction & sample processing, identification and quantification. There are mainly three approaches (selective sampling, bulk sampling and volume reduction) that have been used for sampling of MPs from water, sediment and biota. The collected samples were taken to the laboratory to isolate or extract MPs followed by sample preparation for the required analysis of the MPs technique. It includes filtration, sieving, visual separation, magnetic separation, electrostatic separation, density extraction, *etc*. The isolated heterogeneous or homogeneous sample will be processed according to the characterization technique. The imaging and microscopy (Optical microscopy and SEM) provide morphological information (shape, size and texture) of the MPs surface. The spectroscopic analysis (FTIR and Raman spectroscopy) can identify the polymer based on the chemical composition of the MPs. The majority of the spectroscopic analytical techniques are non-destructive; however, they are time-consuming and provide qualitative information only. In the last few years, it has been reported that various combined analytical techniques have been used such as SEM-FTIR, μ-FTIR, Raman microscopy, GC-MS, Pyr-GC-MS, TGA, DSC and PEE. The emerging technology is still rarely used and is more expensive. Further analytical development is required to figure out the accurate concentration of MPs from the environmental sample. Some of the QA/QC needs to be applied in each step of analysis (from sampling to quantification). A few recommendations have been mentioned in the present chapter to avoid cross-contamination of MPs. The various methods for microplastic identification and their respective advantages and disadvantages might better help a researcher to formulate a pathway for the identification of microplastics.

ACKNOWLEDGMENTS

The authors are grateful and would like to acknowledge DST for providing INSPIRE fellowship to Mr. Gautam Priyadarshi, and to the SHODH scheme by the Gujarat Government for providing fellowship to Mr. Sagar Prajapati during the course of this study.

REFERENCES

[1]   Souza Machado AA, Kloas W, Zarfl C, Hempel S, Rillig MC. Microplastics as an emerging threat to terrestrial ecosystems. Glob Change Biol 2018; 24(4): 1405-16.
[http://dx.doi.org/10.1111/gcb.14020] [PMID: 29245177]

[2]   Dris R, Gasperi J, Mirande C, *et al.* A first overview of textile fibers, including microplastics, in indoor and outdoor environments. Environ Pollut 2017; 221: 453-8.
[http://dx.doi.org/10.1016/j.envpol.2016.12.013] [PMID: 27989388]

[3]   Manzoor S, Naqash N, Rashid G, Singh R. Plastic material degradation and formation of microplastic in the environment: A review. Mater Today Proc 2021; 56(part 6): 3254-60.
[http://dx.doi.org/10.1016/j.matpr.2021.09.379]

[4]   Deng H, Wei R, Luo W, *et al.* Microplastic pollution in water and sediment in a textile industrial area. Environ Pollut 2020; 258: 113658.
[http://dx.doi.org/10.1016/j.envpol.2019.113658] [PMID: 31838382]

[5]   Piehl S, Leibner A, Löder MGJ, *et al.* Identification and quantification of macro- and microplastics on an agricultural farmland. Sci Rep 2018; 8(1): 17950.
[http://dx.doi.org/10.1038/s41598-018-36172-y] [PMID: 30560873]

[6]   Van Cauwenberghe L, Vanreusel A, Mees J, Janssen CR. Microplastic pollution in deep-sea

sediments. Environ Pollut 2013; 182: 495-9.
[http://dx.doi.org/10.1016/j.envpol.2013.08.013] [PMID: 24035457]

[7]   Yousuf A, Naseer M, Naqash N, Singh R. Isolation and identification of microplastic particles from agricultural soil and its detection by fluorescence microscope technique. Think India Journal 2019; 22(16): 3934-49.

[8]   Kim JS, Lee HJ, Kim SK, Kim HJ. Global pattern of microplastics (MPs) in commercial food-grade salts: sea salt as an indicator of seawater MP pollution. Environ Sci Technol 2018; 52(21): 12819-28.
[http://dx.doi.org/10.1021/acs.est.8b04180] [PMID: 30285421]

[9]   EFSA Panel on Contaminants in the Food Chain (CONTAM)Presence of microplastics and nanoplastics in food, with particular focus on seafood. EFSA J 2016; 14(6): e04501.
[http://dx.doi.org/10.2903/j.efsa.2016.4501]

[10]   Yan Z, Chen Y, Bao X, *et al.* Microplastic pollution in an urbanized river affected by water diversion: Combining with active biomonitoring. J Hazard Mater 2021; 417: 126058.
[http://dx.doi.org/10.1016/j.jhazmat.2021.126058] [PMID: 34015710]

[11]   Revel M, Châtel A, Mouneyrac C. Micro(nano)plastics: A threat to human health? Curr Opin Environ Sci Health 2018; 1: 17-23.
[http://dx.doi.org/10.1016/j.coesh.2017.10.003]

[12]   Prata JC, da Costa JP, Lopes I, Duarte AC, Rocha-Santos T. Environmental exposure to microplastics: An overview on possible human health effects. Sci Total Environ 2020; 702: 134455.
[http://dx.doi.org/10.1016/j.scitotenv.2019.134455] [PMID: 31733547]

[13]   Stock F, Kochleus C, Bänsch-Baltruschat B, Brennholt N, Reifferscheid G. Sampling techniques and preparation methods for microplastic analyses in the aquatic environment – A review. Trends Analyt Chem 2019; 113: 84-92.
[http://dx.doi.org/10.1016/j.trac.2019.01.014]

[14]   Crawford CB, Quinn B. Microplastic separation techniques.Microplastic Pollutants 2017; 203-18.
[http://dx.doi.org/10.1016/B978-0-12-809406-8.00009-8]

[15]   Zhang K, Su J, Xiong X, *et al.* Microplastic pollution of lakeshore sediments from remote lakes in Tibet plateau, China. Environ Pollut 2016; 219: 450-5.
[http://dx.doi.org/10.1016/j.envpol.2016.05.048] [PMID: 27238763]

[16]   Zhang K, Su J, Xiong X, *et al.* Microplastic pollution of lakeshore sediments from remote lakes in Tibet plateau, China. Environ Pollut 2016; 219: 450-5.
[http://dx.doi.org/10.1016/j.envpol.2016.05.048] [PMID: 27238763]

[17]   Tsang YY, Mak CW, Liebich C, *et al.* Microplastic pollution in the marine waters and sediments of Hong Kong. Mar Pollut Bull 2017; 115(1-2): 20-8.
[http://dx.doi.org/10.1016/j.marpolbul.2016.11.003] [PMID: 27939688]

[18]   Güven O, Gökdağ K, Jovanović B, Kıdeyş AE. Microplastic litter composition of the Turkish territorial waters of the Mediterranean Sea, and its occurrence in the gastrointestinal tract of fish. Environ Pollut 2017; 223: 286-94.
[http://dx.doi.org/10.1016/j.envpol.2017.01.025] [PMID: 28117186]

[19]   Browne MA, Galloway TS, Thompson RC. Spatial patterns of plastic debris along Estuarine shorelines. Environ Sci Technol 2010; 44(9): 3404-9.
[http://dx.doi.org/10.1021/es903784e] [PMID: 20377170]

[20]   Campanale C, Savino I, Pojar I, Massarelli C, Uricchio VF. A practical overview of methodologies for sampling and analysis of microplastics in riverine environments. Sustainability (Basel) 2020; 12(17): 6755.
[http://dx.doi.org/10.3390/su12176755]

[21]   Manzoor S, Kaur H, Singh R. Existence of Microplastic as Pollutant in Harike Wetland: An Analysis of Plastic Composition and First Report on Ramsar Wetland of India. Curr W Envir 2021.

[http://dx.doi.org/10.12944/CWE.16.1.12]

[22]   Hanvey JS, Lewis PJ, Lavers JL, *et al.* A review of analytical techniques for quantifying microplastics in sediments. Anal Methods 2017; 9(9): 1369-83.
[http://dx.doi.org/10.1039/C6AY02707E]

[23]   Primpke S, Lorenz C, Rascher-Friesenhausen R, Gerdts G. An automated approach for microplastics analysis using focal plane array (FPA) FTIR microscopy and image analysis. Anal Methods 2017; 9(9): 1499-511.
[http://dx.doi.org/10.1039/C6AY02476A]

[24]   KAUR H, SINGH R. Analysis Of Nylon 6 As Microplastic In Harike Wetland By Comparing Its IR Spectra With Virgin Nylon 6 And 6.6. Eur J Mol Clin Med 2020; 7(07): 2020.

[25]   Covernton GA, Pearce CM, Gurney-Smith HJ, *et al.* Size and shape matter: A preliminary analysis of microplastic sampling technique in seawater studies with implications for ecological risk assessment. Sci Total Environ 2019; 667: 124-32.
[http://dx.doi.org/10.1016/j.scitotenv.2019.02.346] [PMID: 30826673]

[26]   Lv L, Yan X, Feng L, *et al.* Challenge for the detection of microplastics in the environment. Water Environ Res 2021; 93(1): 5-15.
[http://dx.doi.org/10.1002/wer.1281] [PMID: 31799785]

[27]   Phuong NN, Fauvelle V, Grenz C, *et al.* Highlights from a review of microplastics in marine sediments. Sci Total Environ 2021; 777: 146225.
[http://dx.doi.org/10.1016/j.scitotenv.2021.146225]

[28]   Hidalgo-Ruz V, Gutow L, Thompson RC, Thiel M. Microplastics in the marine environment: a review of the methods used for identification and quantification. Environ Sci Technol 2012; 46(6): 3060-75.
[http://dx.doi.org/10.1021/es2031505] [PMID: 22321064]

[29]   Felsing S, Kochleus C, Buchinger S, *et al.* A new approach in separating microplastics from environmental samples based on their electrostatic behavior. Environ Pollut 2018; 234: 20-8.
[http://dx.doi.org/10.1016/j.envpol.2017.11.013] [PMID: 29154206]

[30]   Möller JN, Löder MGJ, Laforsch C. Finding microplastics in soils: a review of analytical methods. Environ Sci Technol 2020; 54(4): 2078-90.
[http://dx.doi.org/10.1021/acs.est.9b04618] [PMID: 31999440]

[31]   Enders K, Tagg AS, Labrenz M. Evaluation of electrostatic separation of microplastics from mineral-rich environmental samples. Front Environ Sci 2020; 8: 112.
[http://dx.doi.org/10.3389/fenvs.2020.00112]

[32]   He D, Zhang X, Hu J. Methods for separating microplastics from complex solid matrices: Comparative analysis. J Hazard Mater 2021; 409: 124640.
[http://dx.doi.org/10.1016/j.jhazmat.2020.124640] [PMID: 33246814]

[33]   Grbic J, Nguyen B, Guo E, You JB, Sinton D, Rochman CM. Magnetic extraction of microplastics from environmental samples. Environ Sci Technol Lett 2019; 6(2): 68-72.
[http://dx.doi.org/10.1021/acs.estlett.8b00671]

[34]   Rhein F, Scholl F, Nirschl H. Magnetic seeded filtration for the separation of fine polymer particles from dilute suspensions: Microplastics. Chem Eng Sci 2019; 207: 1278-87.
[http://dx.doi.org/10.1016/j.ces.2019.07.052]

[35]   Ye S, Cheng M, Zeng G, *et al.* Insights into catalytic removal and separation of attached metals from natural-aged microplastics by magnetic biochar activating oxidation process. Water Res 2020; 179: 115876.
[http://dx.doi.org/10.1016/j.watres.2020.115876] [PMID: 32387922]

[36]   Tang Y, Zhang S, Su Y, Wu D, Zhao Y, Xie B. Removal of microplastics from aqueous solutions by magnetic carbon nanotubes. Chem Eng J 2021; 406: 126804.
[http://dx.doi.org/10.1016/j.cej.2020.126804]

[37] Stock F, Kochleus C, Bänsch-Baltruschat B, Brennholt N, Reifferscheid G. Sampling techniques and preparation methods for microplastic analyses in the aquatic environment – A review. Trends Analyt Chem 2019; 113: 84-92.
[http://dx.doi.org/10.1016/j.trac.2019.01.014]

[38] Claessens M, Meester SD, Landuyt LV, Clerck KD, Janssen CR. Occurrence and distribution of microplastics in marine sediments along the Belgian coast. Mar Pollut Bull 2011; 62(10): 2199-204.
[http://dx.doi.org/10.1016/j.marpolbul.2011.06.030] [PMID: 21802098]

[39] Cole M, Lindeque P, Fileman E, *et al.* Microplastic ingestion by zooplankton. Environ Sci Technol 2013; 47(12): 6646-55.
[http://dx.doi.org/10.1021/es400663f] [PMID: 23692270]

[40] Hurley RR, Lusher AL, Olsen M, Nizzetto L. Validation of a method for extracting microplastics from complex, organic-rich, environmental matrices. Environ Sci Technol 2018; 52(13): 7409-17.
[http://dx.doi.org/10.1021/acs.est.8b01517] [PMID: 29886731]

[41] Bergmann M, Gutow L, Klages M. Marine anthropogenic litter. Springer Nature 2015.
[http://dx.doi.org/10.1007/978-3-319-16510-3]

[42] Karlsson TM, Vethaak AD, Almroth BC, *et al.* Screening for microplastics in sediment, water, marine invertebrates and fish: Method development and microplastic accumulation. Mar Pollut Bull 2017; 122(1-2): 403-8.
[http://dx.doi.org/10.1016/j.marpolbul.2017.06.081] [PMID: 28689849]

[43] Leslie HA, Brandsma SH, van Velzen MJM, Vethaak AD. Microplastics en route: Field measurements in the Dutch river delta and Amsterdam canals, wastewater treatment plants, North Sea sediments and biota. Environ Int 2017; 101: 133-42.
[http://dx.doi.org/10.1016/j.envint.2017.01.018] [PMID: 28143645]

[44] Collard F, Gilbert B, Eppe G, Parmentier E, Das K. Detection of anthropogenic particles in fish stomachs: an isolation method adapted to identification by Raman spectroscopy. Arch Environ Contam Toxicol 2015; 69(3): 331-9.
[http://dx.doi.org/10.1007/s00244-015-0221-0] [PMID: 26289815]

[45] Gregory MR. Plastic pellets on New Zealand beaches. Mar Pollut Bull 1977; 8(4): 82-4.
[http://dx.doi.org/10.1016/0025-326X(77)90193-X]

[46] Morét-Ferguson S, Law KL, Proskurowski G, Murphy EK, Peacock EE, Reddy CM. The size, mass, and composition of plastic debris in the western North Atlantic Ocean. Mar Pollut Bull 2010; 60(10): 1873-8.
[http://dx.doi.org/10.1016/j.marpolbul.2010.07.020] [PMID: 20709339]

[47] Doyle MJ, Watson W, Bowlin NM, Sheavly SB. Plastic particles in coastal pelagic ecosystems of the Northeast Pacific ocean. Mar Environ Res 2011; 71(1): 41-52.
[http://dx.doi.org/10.1016/j.marenvres.2010.10.001] [PMID: 21093039]

[48] Shim WJ, Hong SH, Eo SE. Identification methods in microplastic analysis: a review. Anal Methods 2017; 9(9): 1384-91.
[http://dx.doi.org/10.1039/C6AY02558G]

[49] Eriksen M, Mason S, Wilson S, *et al.* Microplastic pollution in the surface waters of the Laurentian Great Lakes. Mar Pollut Bull 2013; 77(1-2): 177-82.
[http://dx.doi.org/10.1016/j.marpolbul.2013.10.007] [PMID: 24449922]

[50] Fries E, Dekiff JH, Willmeyer J, Nuelle MT, Ebert M, Remy D. Identification of polymer types and additives in marine microplastic particles using pyrolysis-GC/MS and scanning electron microscopy. Environ Sci Process Impacts 2013; 15(10): 1949-56.
[http://dx.doi.org/10.1039/c3em00214d] [PMID: 24056666]

[51] Wang W, Wang J. Investigation of microplastics in aquatic environments: An overview of the methods used, from field sampling to laboratory analysis. Trends Analyt Chem 2018; 108: 195-202.

[http://dx.doi.org/10.1016/j.trac.2018.08.026]

[52] Song YK, Hong SH, Jang M, *et al.* A comparison of microscopic and spectroscopic identification methods for analysis of microplastics in environmental samples. Mar Pollut Bull 2015; 93(1-2): 202-9.
[http://dx.doi.org/10.1016/j.marpolbul.2015.01.015] [PMID: 25682567]

[53] Elkhatib D, Oyanedel-Craver V. A critical review of extraction and identification methods of microplastics in wastewater and drinking water. Environ Sci Technol 2020; 54(12): 7037-49.
[http://dx.doi.org/10.1021/acs.est.9b06672] [PMID: 32432459]

[54] Allen V, Kalivas JH, Rodriguez RG. Post-consumer plastic identification using Raman spectroscopy. Appl Spectrosc 1999; 53(6): 672-81.
[http://dx.doi.org/10.1366/0003702991947324]

[55] Cole M, Lindeque P, Halsband C, Galloway TS. Microplastics as contaminants in the marine environment: A review. Mar Pollut Bull 2011; 62(12): 2588-97.
[http://dx.doi.org/10.1016/j.marpolbul.2011.09.025] [PMID: 22001295]

[56] Huppertsberg S, Knepper TP. Instrumental analysis of microplastics—benefits and challenges. Anal Bioanal Chem 2018; 410(25): 6343-52.
[http://dx.doi.org/10.1007/s00216-018-1210-8] [PMID: 29959485]

[57] Vetrimurugan E, Jonathan MP, Sarkar SK, *et al.* Occurrence, distribution and provenance of micro plastics: A large scale quantitative analysis of beach sediments from southeastern coast of South Africa. Sci Total Environ 2020; 746: 141103.
[http://dx.doi.org/10.1016/j.scitotenv.2020.141103] [PMID: 32795758]

[58] Iri AH, Shahrah MHA, Ali AM, *et al.* Optical detection of microplastics in water. Environ Sci Pollut Res Int 2021; 28(45): 63860-6.
[http://dx.doi.org/10.1007/s11356-021-12358-2] [PMID: 33462694]

[59] Majewsky M, Bitter H, Eiche E, Horn H. Determination of microplastic polyethylene (PE) and polypropylene (PP) in environmental samples using thermal analysis (TGA-DSC). Sci Total Environ 2016; 568: 507-11.
[http://dx.doi.org/10.1016/j.scitotenv.2016.06.017] [PMID: 27333470]

[60] Castañeda RA, Avlijas S, Simard MA, Ricciardi A. Microplastic pollution in St. Lawrence River sediments. Can J Fish Aquat Sci 2014; 71(12): 1767-71.
[http://dx.doi.org/10.1139/cjfas-2014-0281]

[61] Peñalver R, Arroyo-Manzanares N, López-García I, Hernández-Córdoba M. An overview of microplastics characterization by thermal analysis. Chemosphere 2020; 242: 125170.
[http://dx.doi.org/10.1016/j.chemosphere.2019.125170] [PMID: 31675574]

[62] Golebiewski J, Galeski A. Thermal stability of nanoclay polypropylene composites by simultaneous DSC and TGA. Compos Sci Technol 2007; 67(15-16): 3442-7.
[http://dx.doi.org/10.1016/j.compscitech.2007.03.007]

[63] Mansa R, Zou S. Thermogravimetric analysis of microplastics: A mini review. Environ Adv 2021; 5: 100117.
[http://dx.doi.org/10.1016/j.envadv.2021.100117]

[64] Fuller S, Gautam A. A procedure for measuring microplastics using pressurized fluid extraction. Environ Sci Technol 2016; 50(11): 5774-80.
[http://dx.doi.org/10.1021/acs.est.6b00816] [PMID: 27172172]

[65] Dierkes G, Lauschke T, Becher S, Schumacher H, Földi C, Ternes T. Quantification of microplastics in environmental samples *via* pressurized liquid extraction and pyrolysis-gas chromatography. Anal Bioanal Chem 2019; 411(26): 6959-68.
[http://dx.doi.org/10.1007/s00216-019-02066-9] [PMID: 31471683]

[66] Nuelle MT, Dekiff JH, Remy D, Fries E. A new analytical approach for monitoring microplastics in marine sediments. Environ Pollut 2014; 184: 161-9.

[http://dx.doi.org/10.1016/j.envpol.2013.07.027] [PMID: 24051349]

[67]   Chae DH, Kim IS, Kim SK, Song YK, Shim WJ. Abundance and distribution characteristics of microplastics in surface seawaters of the Incheon/Kyeonggi coastal region. Arch Environ Contam Toxicol 2015; 69(3): 269-78.
[http://dx.doi.org/10.1007/s00244-015-0173-4] [PMID: 26135299]

[68]   Wang W, Ndungu AW, Li Z, Wang J. Microplastics pollution in inland freshwaters of China: A case study in urban surface waters of Wuhan, China. Sci Total Environ 2017; 575: 1369-74.
[http://dx.doi.org/10.1016/j.scitotenv.2016.09.213] [PMID: 27693147]

[69]   Kirstein IV, Hensel F, Gomiero A, *et al.* Drinking plastics? – Quantification and qualification of microplastics in drinking water distribution systems by µFTIR and Py-GCMS. Water Res 2021; 188: 116519.
[http://dx.doi.org/10.1016/j.watres.2020.116519] [PMID: 33091805]

[70]   Rocha-Santos T, Duarte AC. A critical overview of the analytical approaches to the occurrence, the fate and the behavior of microplastics in the environment. Trends Analyt Chem 2015; 65: 47-53.
[http://dx.doi.org/10.1016/j.trac.2014.10.011]

[71]   Wang X, Bolan N, Tsang DCW, Sarkar B, Bradney L, Li Y. A review of microplastics aggregation in aquatic environment: Influence factors, analytical methods, and environmental implications. J Hazard Mater 2021; 402: 123496.
[http://dx.doi.org/10.1016/j.jhazmat.2020.123496] [PMID: 32717542]

[72]   Fischer M, Scholz-Böttcher BM. Simultaneous trace identification and quantification of common types of microplastics in environmental samples by pyrolysis-gas chromatography–mass spectrometry. Environ Sci Technol 2017; 51(9): 5052-60.
[http://dx.doi.org/10.1021/acs.est.6b06362] [PMID: 28391690]

[73]   Uurasjärvi E. Doctoral dissertation, Itä-Suomen yliopisto. Microplastics: A challenge for environmental analytical chemistry

[74]   Gasperi J, Wright SL, Dris R, *et al.* Microplastics in air: Are we breathing it in? Curr Opin Environ Sci Health 2018; 1: 1-5.
[http://dx.doi.org/10.1016/j.coesh.2017.10.002]

[75]   Torre M, Digka N, Anastasopoulou A, Tsangaris C, Mytilineou C. Anthropogenic microfibres pollution in marine biota. A new and simple methodology to minimize airborne contamination. Mar Pollut Bull 2016; 113(1-2): 55-61.
[http://dx.doi.org/10.1016/j.marpolbul.2016.07.050] [PMID: 27491365]

[76]   Wesch C, Bredimus K, Paulus M, Klein R. Towards the suitable monitoring of ingestion of microplastics by marine biota: A review. Environ Pollut 2016; 218: 1200-8.
[http://dx.doi.org/10.1016/j.envpol.2016.08.076] [PMID: 27593351]

[77]   Roch S, Brinker A. Rapid and efficient method for the detection of microplastic in the gastrointestinal tract of fishes. Environ Sci Technol 2017; 51(8): 4522-30.
[http://dx.doi.org/10.1021/acs.est.7b00364] [PMID: 28358493]

CHAPTER 2

Occurrence and Source of Microplastic in the Environment

Sarabjeet Kaur[1,*] and **Monita Dhiman**[2]

[1] *Mehr Chand Mahajan DAV College for Women, Sector- 36, Chandigarh, India*

[2] *Khalsa College for Women, Civil Lines, Ludhiana, Punjab, India*

Abstract: Microplastics are ubiquitous on the earth, even in the purest environments like arctic snow, inaccessible mountains, *via*. Microplastics may be disseminated *via* air fallout near metropolitan areas, however, the great bulk of data points to water as the primary distribution channel. Researchers have discovered that surface and groundwater are also polluted by microplastics, despite maximum research focusing on marine pollution. The international community visualizes a decline in the concentration of floating plastic waste as an essential step toward the long-term sustainability of the seas. However, there is presently no universally acknowledged indicator of floating plastics trash density. Ultimately, a significant portion of the present microplastic proliferation has been attributed to wastewater, which is frequently not efficiently treated to eliminate such tiny, hydrophobic pollutants. Previously treated wastewater is discharged into water bodies, which in turn feed natural water reserves. Microplastics are also dispersed into the soil and terrestrial ecosystems by certain communities that irrigate their crops with wastewater. A further problem is that micro plastic-rich sludge from wastewater facilities is used as a fertilizer for food crops. It is crucial to keep an eye out for new developments in bioplastics and biodegradable polymers that avoid the build-up of microplastics in the food and agriculture industries.

Keywords: Atmosphere, Bioplastic, Biosphere, Distribution, Freshwater, Hydrosphere, Infiltration, Lithosphere, Marine, Microplastic, Nanoplastic, Occurrence, Plastic footprints, Pollution, Primary source, Secondary source, Sewage, Sources, Ubiquitous, Weathering.

INTRODUCTION

The extensive production of synthetic plastic commenced in 1907. Since then, plastic has become an integral part of global life. Plastics are in great demand because of their adaptability, resistance to corrosion, high strength-to-weight

---

[*] **Corresponding Author Sarabjeet Kaur**: Mehr Chand Mahajan DAV College for Women, Sector- 36, Chandigarh, India, E-mail: jatindersarab@gmail.com

Rahul Singh and Neeta Raj Sharma (Eds.)

ratio, low thermal and electrical conductivity, durability, and cheap production cost. As a result, plastic is used in a wide variety of goods, ranging from paper clips to spaceships. Between 1950 and 2015, around 6 billion tonnes of plastic waste were generated from an estimated 8 billion tonnes of produced plastic. Of this, 79% [1] still remains on the earth occupying terrestrial, freshwater, or marine ecosystems. By 2018, around 380 million tonnes of plastic have been added to the world's oceans [2]. Since the pandemic breakout, the amount of plastic garbage generated globally is expected to have increased to 1.6 million tonnes per day, owing to medical waste leaving plastic environmental footprints [3]. Global plastic production is expected to be bifold within two decades [4]. Approximately 6% of the oil produced in the world is consumed for the synthesis of plastic polymers [5]. Therefore, plastics and greenhouse gas emissions are intricately connected from manufacture to dumping, contributing to climate change [5 - 7]. It has been reported that greenhouse gas emissions from plastics worldwide [8, 9] have recently gained attention because of their deleterious effects on the climate. The greenhouse emission is intended to approach 1.34 gigatons annually by 2030 and 2.8 gigatons annually by 2050. On the other hand, plastic is a relatively recent and rising concern to ecosystems compared to other problems, including climate change, global warming, ocean acidification, land-use patterns, pollution, and invasive species. As a result, modern ecosystems will frequently be confronted with a cascade of accumulated stresses in the future.

The manufacture of synthetic plastics derived from fossil fuels continues to grow, but indigent waste management has resulted in stringent pollution problems. Microplastics have been found in coral reefs [10] marine sediments [11], urban and rural regions [12], freshwaters [13], and seawaters [14]. The majority of findings indicate that microplastics accumulate in aquatic ecosystems and their breakdown by products resulting in increased exposure of existing species to microplastics [15, 16]. A recent study on microplastics has focused primarily on the marine environment. Microplastics' implications, particularly in freshwater and soil, are still in their infancy, and significant gaps persist.

SOURCES

Microplastics infiltrate our ecosystems in a variety of ways and from divergent sources. Emissions from industrial and sewage facilities as well as agricultural land runoff are among the pollutants that can contribute to microplastics in the environment. Certain microplastics are designed exclusively for use as abrasives in industry or in personal care items such as exfoliants. As the name suggests, these are made with purpose and are the first generation of microplastics so-called primary or intentional microplastics [17]. To put it another way, secondary microplastics include those formed by the breakdown of more oversized plastic products, such as automobile tyres and synthetic textiles [18] that wear and tear, or as city dust (a collective term for particles produced by synthetic material linked with metropolitan areas) [19, 20]. Once in the ecosystem, plastics and microplastics may circulate through terrestrial, freshwater, and marine environments in cycles.

Primary Source

Primary microplastics, as described by Cole and colleagues [21], are characterized as plastics made that are tiny in size and derived from industrial and household items [22, 23]. Most often, they are utilized in air blasting media [24], face cleansers and cosmetics, as well as in medicine for drug delivery [25]. However, even though contemporary wastewater treatment practices can eliminate up to 99 percent of microplastics, the amount of microplastics discharged into the environment through effluent remains considerable [26].

A form of primary microplastic that has received the most attention is scrubbers, used in products such as exfoliating hand cleansers and face scrubs to remove dead skin cells. In the 1980s, the usage of microplastic scrubbers rose considerably as cosmetic businesses patented these items. The size, shape, and content of the particles vary depending on the type of cosmetic (Fendall and Sewell, 2009); for example, polyethylene and polypropylene granules (<5 mm) as well as polystyrene spheres (<2 mm) have been discovered in the same cosmetic product. Primary microplastics are explicitly created for their utility in personal care products, such as resin pellets and exfoliants [21, 27, 28].

Microplastics are present in varying size fractions across the world's coastal regions and aquatic ecosystems as a result of transport phenomena such as wind and ocean currents. Microplastics have a positive correlation with human population density [29]. Browne [18] conducted a survey of microplastic pollution around shorelines and discovered that one of the primary origins of microplastics in the oceans was sewage contaminated by fibers from washing clothing [30], as marine sediments were comparable to those used for textiles [29,]. These sources account for 35% of all sources [31].

Worldwide, wastewater treatment projects are also a significant contributor to microplastic discharge [18, 32, 33]. While bulky plastic particles are eliminated effectively during wastewater treatment, microplastics frequently circumvent treatment units [32, 34], infiltrating and accumulating in the aquatic environment [35]. Notably, a significant count of water treatment plants are positioned near the ocean creating a significant source of microplastic emission.

Microplastics were considered to have originated from the materials used to manufacture the bottles and caps. However, significant levels of micro-particles of various sizes have been discovered in bottled waters [36, 37]. Nonetheless, based on the nature of discovered microplastics (containing polymers such as polyethylene, polypropylene, and styrene-butadiene-copolymer), the authors hypothesized that microplastics might originate from sources other than packaging materials [37]. Polymeric packaging [38], has revolutionized the food and agriculture sector adding to the problem. Furthermore, in the food service sector, the widespread use of disposable plastic cutlery, cups, and other serving containers is becoming prevalent [38] due to their tenacity, hydrophobicity, and impediment to degeneration.

The other two significant sources of primary microplastics, accounting for 28% and 24% respectively, are tyre dust from road abrasion and city dust from infrastructure abrasion. Road markings (varnish, thermoplastic, prefabricated polymer tape, and epoxy), maritime coatings (dye, polyurethane, and epoxy), PPCPs (polypropylene copolymers), and plastic pellets collectively contribute a small number of microplastics to the environment. As a result, microplastics are easily transported into aquatic environments, where they can build up over time [39].

Secondary Source

The larger pieces of plastic degrade over time on land and at sea, either mechanically or due to UV light exposure, weakening their structural integrity and enabling an increasing amount of microplastics to enter the ecosystem [40]. Microplastics, primarily found on beaches, are the perfect conditions for plastic fragmentation because they contain both chemical and mechanical weathering [41]. Additionally, it has been proven that 80 percent of marine debris originates on land [42, 43]. This enhances the availability of plastic trash for consumption by a wide range of creatures and focuses on the emergence of additional environmental concerns [39].

Due to their limited susceptibility to breakdown when exposed to seawater and sunshine, single-use plastic bags are a substantial cause of secondary microplastics [44]. Additionally, packing debris, fishing lines, and nets are sources [45]. Disposable consumer products such as plastic floats used in aquaculture practices and docks can get shattered as a result of isopod drilling activity, with every adult isopod producing between 4900 and 6300 microplastic particles during the boring process [46]. These makeup between 69 to 81% of the microplastics detected in the seas.

Microplastics are also a result of the development of biodegradable plastics made of typically fabricated polymers, starch, and vegetable oils and engineered to degrade more quickly [39, 47, 48].

Plastics have been projected to have a lifespan ranging from days to years [40, 49], but this figure is still unknown due to the fact that it has been only for approximately sixty years that the typical plastics have been manufactured and stockpiled.

Occurrence

As previously stated, residual plastic in the environment degrades over time to form microplastics and finally nanoplastics. Microplastics are present in a variety of habitats globally in discrete range fragments due to transport events such as wind, tidal motion and ocean currents.

In order to precisely estimate the amount of microplastic in the environment, we need to explore its origins, storage, and transportation. The amount of microplastics, for example, is based on a precise estimation of 'mismanaged trash,' which is reported in different ways by different nations. Furthermore, the number of possible entry points for plastic and microplastic into the environment is so vast that a precise count is nearly impossible. Monitoring and detecting microplastics after they've been released into the environment can be complex, especially if the particles are deep in the water or extremely tiny.

The most significant concentrations of microplastics in freshwater have been measured in river catchments in the northwest of England, at 517,000 particles per square meter. However, this reflects the current absence of internationally recognized and standardized methodologies for assessing microplastics in the environment. In other regions of the world, river microplastic concentrations are expected to be far more significant.

Plastics are known to be light weighted and buoyant, and a significant chunk of microplastics have been appraised as floating debris when sampling from the superficial water layer in a variety of freshwater bodies, ranging from 1 to 105 particles per square kilometres [50, 51], while the denser plastics can easily immerse [16] and they have been detected in bottom deposits [34, 52]. However, a substantial proportion of plastic molecules tend to float in the water column [34, 53]. According to Yuan [52], particles less than 0.5 mm in diameter were consistently the most numerous (31.2–74.4 percent of the total number). Su [54] and team evaluated the concentration of microplastics (N5m) in Taihu Lake, China. Microplastics (about 10 m–5 mm) concentrations in Amsterdam canal water vary from 48 to 187 particles per litre [34].

Microplastics in the water are estimated to weigh between 93,000 and 236,000 tonnes. These estimates, however, are based on plastic visible on the ocean's surface. Most of the time, the current estimates of plastic and microplastic in freshwater, sediments, and soil are way off. This is because of the complexity and interdependencies of these systems and how they all work together. Microplastic concentrations in freshwater and soil can vary according to rainfall and streamflow patterns, including flooding events.

Additionally, microplastics can be used to transport other harmful compounds [16, 55]. However, a few investigations on the prevalence of microplastics in a variety of meals and packaged beverages have been undertaken [56].

Microplastics in Hydrosphere

More industrialized regions like China, Europe, and North America are the primary focus of research on microplastics in aquatic habitats; nevertheless, Africa and South America remain understudied. First, marine environments [16, 21, 57, 58], then wastewater [43, 59 - 61] and now rivers and lakes [62 - 64] became the focus of microplastics study.

Marine Water

According to Woodall and colleagues [65], marine ecosystems have been recognised as important sinks for microplastic pollution, and as a result, various research works have been conducted in this area recently [66 - 71]. As far back as 1971 researchers found spherules (0.1–2.0mm) in the southern New England coasts, spherules (0.2–4.9 mm) in the surface waters of the Atlantic Ocean and pellets (2.5–5.0mm) on the surface of Sargasso Sea [72, 73]. There has been extensive research looking at microplastics in the ocean since Thomicroplasticson's discovery of their presence in sand collected from UK beaches and estuaries in 2004 [74]. These studies have included shorelines from every continent [18, 75], islands large and small [76, 77], and waterways from the Atlantic [78, 79] to the Pacific [80] and Arctic [81] oceans.

Marine microplastic research is classified into two types: peripheral water research (coastlines and sea glaciers) and seafloor studies (deep sea investigations). This division develops due to the fact that various forms of plastic have varying compactness. The consistency of each polymer group is governed by its individual molecular design and crystallinity, with close-bound particles ensuing larger densities, resulting in certain microplastics to be less compact than saltwater (*e.g.* polypropylene) while others are more compact than seawater (*e.g.* acrylic) [65]. On shorelines and the surface of the water, less dense microplastics aggregate, whereas denser particles accumulate in deeper layers of water [18, 82].

According to Isobe [83] and Cózar [84], when the particle size of a big plastic particle shrinks, its volume and weight remain the same, resulting in an exponential increase in the number of microplastics. As a result, particle size affects both the total number of particles and the concentration of microplastics in all of the measurement units. For example, if only large particle sizes are measured, a low level of contamination may appear, but when smaller particle sizes are sampled and examined, a higher level may be shown. Microplastic abundance in the North Atlantic gyre has not changed in over a decade, but Claessens [85] found a rise in the quantity of microplastic in sediment tested from around the Belgian coast during the same period of time (55 to 156 particles/kg of dry sediment). Many studies have shown low-density microplastics to be primarily found in the superficial microlayer of the sea [47], but their distribution in the water column can vary, for example in estuaries, and they can descend if marine fauna adheres to them [47, 40], as has been shown in several studies.

Seawater plastic particles that are less compact than seawater (about 1.02 g/cm3 at 20°C, 1.00-3.0 at 0°C) are gathered, carried, and concentrated by the wind and other oceanic tides [69]. After a storm, microplastics can be suspended on the seafloor and spread throughout the water column [86]. This litter may be found in distant places such as the poles of the earth [87], deep oceans [65] and islands [88] due to the fact that ocean currents [76, 86], wind outflows, and drift all contribute to the distribution of microplastic waste across enormous distances [89]. As a result, microplastics are dispersed throughout marine habitats and are concentrated in the five big trash patches situated in the Atlantic Ocean and Pacific Ocean [90]. While attachment to other materials (such as microorganisms, minerals, biofilms, and faecal matter) can alter the density of polymers and cause them to sink or float at varied rates, it is possible that this attachment will cause less dense microplastics to sink or denser microplastics to float at the ocean's surface [62]. Because of their lower densities, polypropylene (PP) and polyethylene (PE) make up the majority of the microplastics discovered on the ocean's surface [90]. Samples of microplastics collected from 18 shores around the world, including Port Douglas and Busselton Beach (Australia), Kyushu (Japan), Dubai (UAE), Vina Del Mar (Chile), Western Cape (South Africa), Virginia and California (USA), and Sennon Cove (United Kingdom), contained a majority of polyester (56 percent), followed by acrylic (23 percent), polypropylene (7 percent), polyethylene (6 percent), and polyamide fibres (3 percent) [18]. As a result, the types of microplastics found in ocean surface waters (rubbish patches) and along coastlines vary widely. Diverse polymer types travel differently, which might explain some of these discrepancies, while different sources impacting ocean water and shorelines could explain others. In order to correctly explain this phenomenon, further investigation is required to define the origins of plastics and to determine if various polymer types follow discrete transit paths.

A tiny proportion of microplastics detected in the seas are associated with activities such as marine fishing (which uses plastic apparatus), whereas the majority proportion (approximately 80%) includes plastic waste produced from land [57]. The primary contributors to the marine compartment's microplastic contamination include littered beaches and coastal regions [91], terrestrial rivers [92], stormwater runoff [89] marine species passive absorption [93, 94], wastewater discharge, and air microplastic deposition.

Microplastics are more prevalent in estuaries than in seawater samples [95] as a result of anthropogenic trash entering freshwater systems, and beaches and one being washed back by surface currents arising in the ocean [96]. Numerous research on estuaries have been conducted with the goal of demonstrating the existence of microplastics in superficial waters [97] and deposits [98], primarily on coastal beaches [99]. This high level of microplastic contamination in estuaries implies that rivers from land are a significant cause of microplastic effluence in coastal and marine habitats [100].

Fresh Water

Numerous investigations [101, 102] have established that the convergence and dissemination of microplastics on the surface of the water, in the water column, and in the sediment are dependent on parameters such as topographical location, wind, water currents, and rate of flow of stream [103]. Microplastic concentrations in freshwater sources are often higher in metropolitan areas than in rural areas [102]. Additionally, the rate of flow can affect the incidence, clustering, and migration of microplastics in surface freshwater systems [102]. Tibbetts *et al.* [102] ascribed the elevated concentrations to diminished flow pace as the water arrived the lake, resulting in an increase in fine sediment and microplastic deposition.

There were fewer than one microplastic particle per meter square at the Rhine River's surface [104], and seven particles per cubic meter along the Ofanto River in Southeast Italy [105], while higher proportions were detected along the Snake and Lower Columbia rivers [106] and in the superficial waters of northwestern China's Wei River Basin [107]. The Great Lakes Basin of North America is one of the most researched lakes [108], with an average plethora of microplastics floating on the surface of up to 0.043 particles/m2 [109]. In Europe, a comparable concentration (0.048 particles/m2) was recorded in Lake Geneva, Switzerland [110]. Similar studies found that lakes, dams, floodplains, and other low-velocity habitats such as meander cut-offs have greater microplastic concentrations than rivers and streams [111 - 114].

Increased flow velocity and amount have also been associated with accelerated microplastic transit, with floods (enhanced flow pace and volume) producing microplastic flushing in river networks [115]. In general, lakes and low-flow locations function as microplastic sinks, whereas rivers and serve as microplastic transportation systems. As a result, the loading procedures for microplastics are analogous to those for other contaminants [116, 114]. On the other hand, lakes and rivers may be significant sources of secondary microplastics [103, 117, 118]. As in marine habitats, its production may be traced back to the breakdown of fragile plastic debris by currents, streams, and waves [103, 114].

Microplastics can have a deleterious influence on aquatic organisms in freshwater ecosystems. The incidence of microplastics in the digestive system of fish can result in barriers that cause visible damage or histopathological changes in the intestine [42], behavioral and lipid metabolic changes, and transfer to the liver [119]. Microplastics are easily absorbed by organisms (fauna and flora) and then transferred into food webs taking advantage of their tiny size and slow bio-degeneration rate [120]. The presence of microplastics has significantly altered the microbial community in aquatic habitats, which has led to an increase in pathogenic organisms and plastic decomposition organisms [121]. However, these findings require further investigation to establish their validity and to determine the influence of microplastic-induced changes in microbial communities on human health and the environment.

Microplastics in Lithosphere

Terrestrial ecosystems are estimated to receive 4-23 times more microplastic waste annually [122] as compared to aquatic ecosystems. Microplastics' influence on terrestrial habitats is currently poorly understood, owing to a dearth of prior research [123]. One reason for the paucity of prior research is the adversity of isolating and measuring plastic molecules from soil using the latest available equipment, as well as the absence of standardized separation, extraction, and analytical methodologies [124, 125]. Soil settings encourage the growth of biofilms on microplastic particles, which can absorb contaminants such as mineral particles and organic debris prior to further examination. Other pollutants are difficult to remove for three reasons. To begin, some polymers' reactivity to the strong acids or alkalis utilized in digesting procedures [126]. Second, the prohibitively high expense of enzymatic digestion (which is ineffective at eliminating stable organic materials) on large sample sizes [126]. Finally, the possibility of unintentional removal of associated contaminants (*e.g.* plasticizers) or modification of microplastics by wet oxidation is unknown [126]. Numerous attempts have been made to increase the efficiency with which microplastics are extracted from soils. According to Li *et al.* [127], performing pre-digestion with 30% hydrogen peroxide at 70°C increased microplastic extraction from soil and sludge.

Polyethylene and polypropylene were found to be the most prevalent polymers in soil settings, followed by polyvinyl chloride and polyethylene terephthalate [126]. Often, the majority of microplastic particle types detected in soils are composed of fibres obtained from biosolids or treated wastewater used for irrigation [128]. The predominant form of plastic molecule found in a particular area's soils is determined by three factors: the predominant microplastic transport mechanisms [102, 129], the distance between origin and sink [61], and the class of polymer generated and liberated at the source [126]. Due to the topographic variability of soil microplastics and the typically longer detainment of soil plastic molecules [130], soil-based microplastics research requires more specific data pertaining to its location than atmospheric microplastic related work.

Microplastics contaminating soils originate from a variety of sources that vary regionally and according to land use policies. Direct microplastic sources in agricultural land use regions will mostly include plastic mulch films, municipal garbage (compost), biosolids (sewage sludge and anaerobic digestate), plastic-coated fertilizers, and atmospheric deposition [120, 128, 130, 131]. A direct key source of microplastics in agricultural soils is plastic mulching, which is utilized to boost agricultural yields by enhancing water and nutrient efficiency, minimizing erosion, lowering disease burdens, and facilitating early planting and/or harvest cropping [132]. Due to the thin nature of plastic mulching films (about 5-20 m thick), extraction from soils is challenging and economically prohibitive [126]. Furthermore, as a result of soil tillage, ultraviolet radiation [133], physical abrasion (abiotic) in the soil column, and biodegradation [134], plastic mulching films break down into microplastics [135]. Bio solids in agricultural areas can include up to 70% to 95% of the microplastics found in household wastewater retained in sludge throughout water treatment [136]. The concentrations of microplastics in sludge imply that significant amounts of microplastics are transmitted to soil ensuing sludge or biosolid application [130]. In non-agricultural environments, various sources of plastics can contaminate soils and groundwater. The landfills and leachate from landfills can include microplastic concentrations of up to 25 particles per litre [137]. Finally, it appears as though point sources dominate soil microplastic contamination. However, in order to design future effective mitigation techniques, it is necessary to understand how these particles migrate inside the soil.

A recent study indicates that the amount and concentration of microplastic often decrease as soil depth increases [128]. Bioturbation [120], soil tillage [128], and water infiltration, including recurrent wet-dry cycles [138], all contribute to the vertical movement of microplastics across soils. The vertical microplastic transfer will thus be accelerated in places with many macro pores and the agro-based areas with artificial irrigation facilities [126]. Microplastics may interact with microbes and macromolecules in the soil profile *via* homo- and hetero aggregation. The size, mold, and chemical configuration of the soil sediments may have an impact on these interactions and the motility of microplastics in soil [125]. Additionally, microplastics can be eroded from or deposited on soils as a result of wind erosion, surface runoff and during the harvesting of crops [61]. Furthermore, animals can transfer and disseminate microplastics on soils by extraneous adherence of plastic particles to the animals or by ingesting and excreting the plastic particles [139, 140]. However, the transport and movement of microplastics in soils are rarely recorded, and the factors that influence these rates are not well understood [125, 141].

On a worldwide scale, however, given the continuous manufacturing and discharge of plastic goods into the ecosystem, as well as the difficulty of removing microplastics from soils and sluggish degradation rates, the progressive build-up of microplastics in soils is unavoidable [142, 143]. Microplastic pollution is manifestly influenced by the origin and routes of synthetic polymer molecules in the soil, as well as by their possible effects [144]. The physical and chemical attributes of the soil, microbial activities, enzyme action, and plant growth are all affected by microplastics They also pose detrimental eco-toxicological outcomes to soil fauna [145]. Therefore, in order to appreciate the hazards connected with soil microplastic contamination and to identify efficient remediation procedures, sources and routes must be explored.

Microplastics in Atmosphere

Microplastics were first discovered to be transported by air and atmosphere in 2015 in Paris [146]. In the study of microplastic contamination in metropolitan areas by Dris *et al.* [146], it was suggested that air fallout can be a major contributor of fibres in freshwater ecosystems. In the current scientific landscape, aerial and atmospheric delivery remain poorly understood methods of delivering MP [147, 148].

Microplastics have been identified in the atmospheres of urban, suburban, and distant places, with significant variation in amount recorded across research locations [61, 149]. Among European cities, Paris and Hamburg, the average wet and dry deposition was 118 (Paris) and 275 (Hamburg) particles per square meter per day, respectively [116, 146, 150]. On the other hand, in Dongguan, China, the non-fibrous microplastics and fibres varying from 175 to 313 particles m2 dl [151] were deposited as atmospheric microplastic.

Observations of atmospheric microplastics from urban areas to distant places confirm that microplastic contamination is a universal issue and that atmospheric microplastic conveyance is a primary route for microplastics reaching distant regions [152, 153]. There is a noticeable rise in the quantity of atmospheric microplastic bits from Paris (0.3-1.5 particles m3) to Shanghai (0 to 4.18 particles m3) [128]. Anthropogenic activities, industrialisation, and an increase in population density are all to blame for this rise [61].

Microplastic abundance in the atmosphere also varies among indoor and outdoor contexts, with the latter often having reduced levels of abundance [61, 157]. For instance, in Paris, outdoor concentrations were detected between 0.3 and 1.5 fibers per cubic meter 331, while interior concentrations ranged between 1.0 and 60.0 fibres per cubic meter [154]. An average concentration of 26,800 mg kg-1 of microplastic was found in indoor dust samples from 39 cities in China, extending from 1550 to 120,000 mg kg-1 [128]. As a result, indoor pollutants are a significant source of atmospheric pollutants, readily devoting to atmospheric deposition [155, 156].

Microplastics may be deposited in the atmosphere as a result of precipitation events in the form of rain and snow [129, 157]. According to Dris *et al.* (2016), microplastic fallout was found to be between 2 and 24 particles per square meter per day on those days when rainfall was as low as 0.02% and 11 to 340 particles per square meter per day on days when rainfall was as high as 2 to 5 millimetres. Recent research has shown that microplastics and nanoplastics can serve as well-organized cloud ice nuclei, which may determine the association between microplastic deposition and snowfall episodes [157]. Ice and snow contribute to the deposition of plastic particles in both urban and distant (such as the Arctic) terrestrial locations, as well as in the world's oceans, where it accumulates as particulate matter and is removed by deposition [61]. Bergman *et al.* [158] observed microplastics assemblage from European and Arctic snow specimens as 190-154 and 0-14.4 103 particles L-1, respectively, due to snow-directed accumulation, which is consistent with previous findings. Rain and snow are both regarded as excellent scavenging mechanisms for aerosol particles based on the results of research exploring precipitation-driven MP deposition [61].

Recent research indicates that wind activity has an effect on the mobility of synthetic polymer molecules and the enhancement of wind-corroded sediments with microplastics. Rezaei [159] showed that less-density microplastics (LDMP) concentrations were higher in wind-eroded sediments (20.27 mg/kg) than in natural soils (6.91 mg/kg) in microplastic-polluted locations in Iran's Fars Province. Additionally, the study found an enrichment ratio of 2.83 to 7.63 for LDMP and a corrosion rate of 0.08 to 1.48 mg m2 min1 [159]. The findings of this study together emphasize the importance of wind erosion in the expansion of microplastics, which may pose a vulnerable threat to people by direct inspiration of the particles [159]. To estimate and anticipate the movement of microplastics, scientists have begun leveraging long-term data on wind direction and strength trends. For instance, Genc *et al.* [160] utilized a blend of wind and wave data to recognize places prone to microplastic build-up in Turkey's Fethiye-Göcek Specially Protected Area. Future research and moderation efforts might be targeted using prototypes like the one developed by Genc [160] and others. The inquiry of the worldwide microplastic cycle may benefit greatly from the use of these models when they have been fully tested and validated for correctness.

Potentially, the particle size and form of microplastics might affect their air transit [61]. Particles moved through the atmosphere are often lower in size in comparison to aquatic and terrestrial microplastics. Roughly half of the air microplastic fibers examined in the Pyrenees Mountains were less than 300 meters in length, and 70% of microplastic pieces were less than 50 meters in length [157]. European and Arctic microplastics, on the other hand, contribute 80% and 98% of the overall quantity of microplastics in the snow, respectively [158]. These findings reveal that the quantity of microplastics in the atmosphere reduced as particle size increased [158], implying that particle dimension is an important aspect in the atmospheric transport of microplastics. On the other hand, the influence of MP shape is less well known, despite the fact that a variety of forms have been detected in air samples [61]. Atmospheric microplastics come in a range of forms, from fibers to films, foams, and fragments, and their origins range from tyre abrasion [161] to weathered macroplastics [61]) and other regional land-based sources [162]. However, several investigations have discovered that the predominant forms of atmospheric microplastics vary with location. For example, in Dongguan, Shanghai, Yantai, and Paris, fibers were the predominant shape (>60%) of atmospheric microplastics discovered, but in Hamburg, fragments were the predominant form (95%) of the average microplastic deposition [150]. In Iran, on the other hand, microplastics found in street dirt were primarily granules (65.9 percent) and fibers (33.5 percent) [163]. Different microplastic shapes may influence how these particles travel through their surroundings, including the atmosphere. For instance, the enhanced surface area of thin, flat films might facilitate air transportation in comparison to pieces of comparable mass [157]. Around 34% of automotive tyre wear particles were deposited in the seas by air transport, which is close to direct and riverine transference estimates [153]. Following that, the effect of form on atmospheric microplastic transport requires more exploration.

Microplastics in Biosphere

Due to their tiny size, microplastics are swiftly absorbed by plants and animals [164] and induce a variety of fitness concerns stretching from altered gene expression and behaviour to decreased reproductive productivity [165, 166].

This intrinsic toxicity is further exacerbated by the fact that microplastics have a huge, extremely hydrophobic surface area, making them excellent absorbers of endocrine disruptors and other environmental toxins [167, 168].

There is evidence that microplastics can be absorbed by seafood [169] and crop plants like (*Triticum aestivum*) and lettuce (*Lactucasativa*) through fissures at the developing locations of fresh lateral roots [63]. Following their entry into roots, microplastics are then transferred to shoots [63]. Plant transpiration drives the rate at which microplastics are absorbed, with higher rates of transpiration resulting in more absorption [63]. As a result, the physiological and morphological properties of various plant species may have an effect on microplastic absorption.

Other variables affecting microplastics' uptake and possible damage to terrestrial wildlife and flora include the particle's size and form, with smaller angular particles easily flowing through membrane obstacles than particles with ordered surfaces or stretched edges [170]. The kind of polymer also has a noteworthy role in defining the effects of microplastics on organisms. Polyethylene microplastics, for example, had a less detrimental effect on wheat development than microplastics resulting from starch-based plastic mulching film [171]. There is still a dearth of understanding of the particular mechanisms that lead to polymers in the lithosphere, especially in land-dwelling species, causing deleterious impacts.

Increased mortality and lower growth rates were seen in terrestrial fauna when microplastics surpassing 28% dry weight of soil were ingested by earthworm populations [172]. Additional studies found that microplastic concentrations as low as 1% and 2% dramatically reduced the development of earthworms and augmented their mortality, respectively [139]. Microplastics primarily injure soil-dwelling species by accumulating plastic particles inside their digestive tracts, compromising their immune systems and impairing feeding behaviour and growth [126]. Thus, in terrestrial contexts, the consumption of microplastics by earthworms, mites, *etc.*, which are critical for maintaining soil quality, might possibly result in decreasing soil quality as mesofauna abundance declines. As a consequence, microplastics may signify a hazard to terrestrial species and to the long-term viability of agroecosystems [126, 173].

Microplastic ingestion by humans has yet to be proven hazardous, although bioaccumulation patterns indicate it might have a deleterious impact on human health [174, 175]. Human beings would be exposed to airborne microplastics *via* inspiration and dust consumption, triggering potential hostile effects on human well-being [59]. This can cause inflammation and secondary Geno toxicity [176]. The airborne microplastics are associated with unreacted monomers, additives of plastics, and other destructive pollutants engrossed from the environments (such as polycyclic aromatic hydrocarbons, persistent organic pollutants, heavy metals and microbes). These vicious compounds may boost the toxicity of aerial microplastics to human beings.

CONCLUSION

An increasing danger to food and farming is posed by microplastics made from polymeric materials derived from petroleum. All of these factors, including their toxicity, adsorption capacity, and resistance to degradation, have a detrimental effect on ecosystems ranging from seas to soils to insect digestive tracts [142]. Several procedures for curtailing microplastic contamination [177] have been proposed, falling essentially into three categories: containment, mitigation, and separation. The goal of containment is to use proper plastic disposal techniques such as recycling and well-maintained landfills [178], to mitigate human conduct that contributes to microplastic pollution, and to separate microplastics from wastewater during processing. As a consequence of mounting evidence that microplastics harm the environment, several countries have debarred the usage of plastic microbeads in cosmetics, a significant cause of key microplastics [179]. Stricter litter rules also limit the amount of garbage and microplastics that are disposed of in an incorrect manner [179]. Furthermore, some governments have supported education initiatives to promote proper waste management practices, such as recycling [180] and using garbage containers [179]. While supplementing present wastewater treatment actions with enzymatic microplastic decomposition may answer fears about exclusion efficacy and recontamination avoidance, practical issues remain. Biodegradation by enzymology has been devised as a possible solution to this disruption.

It is crucial to keep an eye out for new developments in bioplastics and biodegradable polymers that avoid the build-up of microplastics in food and agriculture productions. We urge that the food and agricultural systems invest significantly in biodegradation. The number of domestic and transnational promises to minimize microplastic pollution is increasing as global plastic output continues to climb. National governments are increasingly prohibiting or taxing single-use plastic items [181, 182]. The United Nations (UN) has established worldwide pledges to decrease plastic leaks into the environment. The UN Environment Assembly Resolution on Marine Litter and Microplastics, as well as the Sustainable Development Goals (SDGs) are among these commitments.

ACKNOWLEDGEMENTS

The authors wish to acknowledge the researchers who have contributed to generating information during different periods through their research which helped in the compilation of the document.

REFERENCES

[1] Geyer R, Jambeck JR, Law KL. Production, use, and fate of all plastics ever made. Sci Adv 2017; 3(7): e1700782.
[http://dx.doi.org/10.1126/sciadv.1700782] [PMID: 28776036]

[2] Ritchie H, Roser M. Plastic pollution. Our World in Data 2018.

[3] Benson NU, Bassey DE, Palanisami T. COVID pollution: impact of COVID-19 pandemic on global plastic waste footprint. Heliyon 2021; 7(2): e06343.
[PMID: 33655084]

[4] Lebreton L, Andrady A. Future scenarios of global plastic waste generation and disposal. Palgrave Commun 2019; 5: 1-11.

[5] Zhu X. The plastic cycle – an unknown branch of the carbon cycle. Front Mar Sci 2021; 7: 1227.
[http://dx.doi.org/10.3389/fmars.2020.609243]

[6] Benavides PT, Lee U, Zarè-Mehrjerdi O. Life cycle greenhouse gas emissions and energy use of polylactic acid, bio-derived polyethylene, and fossil-derived polyethylene. J Clean Prod 2020; 277: 124010.

[7] Walker TR, McKay DC. Comment on five misperceptions surrounding the environmental impacts of single-use plastic. Environ Sci Technol 2021; 55(2): 1339-40.
[http://dx.doi.org/10.1021/acs.est.0c07842] [PMID: 33389994]

[8] Shen M, Huang W, Chen M, Song B, Zeng G, Zhang Y. (Micro) plastic crisis: un-ignorable contribution to global greenhouse gas emissions and climate change. J Clean Prod 2020; 254: 120138.
[http://dx.doi.org/10.1016/j.jclepro.2020.120138]

[9] Rahimi A, García JM. Chemical recycling of waste plastics for new materials production. Nat Rev Chem 2017; 1: 1-1.

[10] Cordova M R, Hadi T A, Prayudha B. Occurrence and abundance of microplastics in coral reef sediment: a case study in Sekotong, Lombok-Indonesia. Microplastics 2018; 10(1): 23-9.
[http://dx.doi.org/10.5281/zenodo.1297719]

[11] Van Cauwenberghe L, Vanreusel A, Mees J, Janssen CR. Microplastic pollution in deep-sea sediments. Environ Pollut 2013; 182: 495-9.
[PMID: 24035457]

[12] Hirai H, Takada H, Ogata Y, *et al.* Organic micropollutants in marine plastics debris from the open ocean and remote and urban beaches. Mar Pollut Bull 2011; 62(8): 1683-92.
[PMID: 21719036]

[13] Faure F, Demars C, Wieser O, Kunz M, De Alencastro LF. Plastic pollution in Swiss surface waters: nature and concentrations, interaction with pollutants. Environ Chem 2015; 12(5): 582-91.

[http://dx.doi.org/10.1071/EN14218]

[14]   Law KL, Thompson RC. Oceans. Microplastics in the seas. Science 2014; 345(6193): 144-5.
[PMID: 25013051]

[15]   Sun J, Dai X, Wang Q, van Loosdrecht MCM, Ni BJ. Microplastics in wastewater treatment plants: Detection, occurrence and removal. Water Res 2019; 152: 21-37.
[PMID: 30660095]

[16]   Andrady AL. Microplastics in the marine environment. Mar Pollut Bull 2011; 62(8): 1596-605.
[PMID: 21742351]

[17]   Ladewig SM, Bao S, Chow AT. Natural fibers. a missing link to chemical pollution dispersion in aquatic environments. Environ Sci Technol 2015; 49(21): 12609-10.
[PMID: 26496674]

[18]   Browne MA, Crump P, Niven SJ, *et al.* Accumulation of microplastic on shorelines woldwide: sources and sinks. Environ Sci Technol 2011; 45(21): 9175-9.
[PMID: 21894925]

[19]   Hidalgo-Ruz V, Gutow L, Thompson RC, Thiel M. Microplastics in the marine environment: a review of the methods used for identification and quantification. Environ Sci Technol 2012; 46(6): 3060-75.
[PMID: 22321064]

[20]   Cole M, Lindeque P, Fileman E, *et al.* Microplastic ingestion by zooplankton. Environ Sci Technol 2013; 47(12): 6646-55.
[PMID: 23692270]

[21]   Cole M, Lindeque P, Halsband C, Galloway TS. Microplastics as contaminants in the marine environment: a review. Mar Pollut Bull 2011; 62(12): 2588-97.
[PMID: 22001295]

[22]   Betts K. Why small plastic particles may pose a big problem in the oceans. Environ Sci Technol 2008; 42(24): 8995.
[PMID: 19174862]

[23]   Moore CJ. Synthetic polymers in the marine environment: a rapidly increasing, long-term threat. Environ Res 2008; 108(2): 131-9.
[PMID: 18949831]

[24]   Jiang J Q. Occurrence of microplastics and its pollution in the environment: A review.Sustainable production and consumption 2018; 13: 16-23.
[http://dx.doi.org/10.1016/j.spc.2017.11.003]

[25]   Patel MM, Goyal BR, Bhadada SV, Bhatt JS, Amin AF. Getting into the brain: approaches to enhance brain drug delivery. CNS Drugs 2009; 23(1): 35-58.
[PMID: 19062774]

[26]   Rochman CM, Tahir A, Williams SL, *et al.* Anthropogenic debris in seafood: Plastic debris and fibers from textiles in fish and bivalves sold for human consumption. Sci Rep 2015; 5(1): 14340.
[PMID: 26399762]

[27]   da Costa JP, Santos PSM, Duarte AC, Rocha-Santos T. (Nano)plastics in the environment - Sources, fates and effects. Sci Total Environ 2016; 566-567: 15-26.
[PMID: 27213666]

[28]   Auta HS, Emenike CU, Fauziah SH. Distribution and importance of microplastics in the marine environment: A review of the sources, fate, effects, and potential solutions. Environ Int 2017; 102: 165-76.
[PMID: 28284818]

[29]   Oerlikon The Fiber Year 2008/09: A world-survey on textile and nonwovens industry. Switzerland: Oerlikon 2009.

[30]   Porter A, Lyons BP, Galloway TS, Lewis C. Role of marine snows in microplastic fate and bioavailability. Environ Sci Technol 2018; 52(12): 7111-9.
[PMID: 29782157]

[31]   Lobelle D, Cunliffe M. Early microbial biofilm formation on marine plastic debris. Mar Pollut Bull 2011; 62(1): 197-200.
[PMID: 21093883]

[32]   Mintenig SM, Löder MGJ, Primpke S, Gerdts G. Low numbers of microplastics detected in drinking water from ground water sources. Sci Total Environ 2019; 648: 631-5.
[PMID: 30121540]

[33]   Long Z, Pan Z, Wang W, *et al.* Microplastic abundance, characteristics, and removal in wastewater treatment plants in a coastal city of China. Water Res 2019; 155: 255-65.
[PMID: 30852313]

[34]   Leslie HA, Brandsma SH, van Velzen MJ, Vethaak AD. Microplastics en route: field measurements in the dutch river delta and amsterdam canals, wastewater treatment plants, north sea sediments and biota. Environ Int 2017; 101: 133-42.
[PMID: 28143645]

[35]   Murphy F, Ewins C, Carbonnier F, Quinn B. Wastewater treatment works (WwTW) as a source of microplastics in the aquatic environment. Environ Sci Technol 2016; 50(11): 5800-8.
[PMID: 27191224]

[36]   Schymanski D, Goldbeck C, Humpf HU, Fürst P. Analysis of microplastics in water by micro-Raman spectroscopy: Release of plastic particles from different packaging into mineral water. Water Res 2018; 129: 154-62.
[PMID: 29145085]

[37]   Oßmann BE, Sarau G, Holtmannspötter H, Pischetsrieder M, Christiansen SH, Dicke W. Small-sized microplastics and pigmented particles in bottled mineral water. Water Res 2018; 141: 307-16.
[PMID: 29803096]

[38]   Sobhani Z, Lei Y, Tang Y, *et al.* Microplastics generated when opening plastic packaging. Sci Rep 2020; 10(1): 4841.
[PMID: 32193409]

[39]   Thomicroplasticson RC. Microplastics in the marine environment: sources, consequences and solutions.Marine anthropogenic litter 2015; pp. 185-2000.

[40]   Barnes DK, Galgani F, Thomicroplasticson RC, Barlaz M. Accumulation and fragmentation of plastic debris in global environments. Philosophical transactions of the royal society B. Biomed Sci 2009; 364(1526): 1985-98.

[41]   Corcoran PL, Biesinger MC, Grifi M. Plastics and beaches: a degrading relationship. Mar Pollut Bull 2009; 58(1): 80-4.
[PMID: 18834997]

[42]   Zhang H. Transport of microplastics in coastal seas. Estuar Coast Shelf Sci 2017; 199: 74-86.

[43]   Enfrin M, Dumée LF, Lee J. Nano/microplastics in water and wastewater treatment processes - Origin, impact and potential solutions. Water Res 2019; 161: 621-38.
[PMID: 31254888]

[44]   Kolandhasamy P, Su L, Li J, Qu X, Jabeen K, Shi H. Adherence of microplastics to soft tissue of mussels: A novel way to uptake microplastics beyond ingestion. Sci Total Environ 2018; 610-611: 635-40.
[PMID: 28822931]

[45]   Welden NA, Cowie PR. Degradation of common polymer ropes in a sublittoral marine environment. Mar Pollut Bull 2017; 118(1-2): 248-53.

[PMID: 28267994]

[46] Davidson TM. Boring crustaceans damage polystyrene floats under docks polluting marine waters with microplastic. Mar Pollut Bull 2012; 64(9): 1821-8.
[PMID: 22763283]

[47] Derraik JG. The pollution of the marine environment by plastic debris: a review. Mar Pollut Bull 2002; 44(9): 842-52.
[PMID: 12405208]

[48] O'Brine T, Thompson RC. Degradation of plastic carrier bags in the marine environment. Mar Pollut Bull 2010; 60(12): 2279-83.
[PMID: 20961585]

[49] Zheng Y, Yanful EK, Bassi AS. A review of plastic waste biodegradation. Crit Rev Biotechnol 2005; 25(4): 243-50.
[PMID: 16419620]

[50] Free CM, Jensen OP, Mason SA, Eriksen M, Williamson NJ, Boldgiv B. High-levels of microplastic pollution in a large, remote, mountain lake. Mar Pollut Bull 2014; 85(1): 156-63.
[PMID: 24973278]

[51] Anderson PJ, Warrack S, Langen V, Challis JK, Hanson ML, Rennie MD. Microplastic contamination in Lake Winnipeg, Canada. Environ Pollut 2017; 225: 223-31.
[PMID: 28376390]

[52] Yuan W, Liu X, Wang W, Di M, Wang J. Microplastic abundance, distribution and composition in water, sediments, and wild fish from Poyang Lake, China. Ecotoxicol Environ Saf 2019; 170: 180-7.
[PMID: 30529617]

[53] Ghosh P, Patra R, Patra P, *et al.* Emerging Threats of Microplastic Contaminant in Freshwater Environment. 2021; pp. 247-58.

[54] Su L, Xue Y, Li L, *et al.* Microplastics in taihu lake, China. Environ Pollut 2016; 216: 711-9.
[PMID: 27381875]

[55] Eerkes-Medrano D, Thompson RC, Aldridge DC. Microplastics in freshwater systems: a review of the emerging threats, identification of knowledge gaps and prioritisation of research needs. Water Res 2015; 75: 63-82.
[PMID: 25746963]

[56] Liebezeit G, Liebezeit E. Synthetic particles as contaminants in German beers. Food Addit Contam Part A Chem Anal Control Expo Risk Assess 2014; 31(9): 1574-8.
[PMID: 25056358]

[57] Li WC, Tse HF, Fok L. Plastic waste in the marine environment: A review of sources, occurrence and effects. Sci Total Environ 2016; 566-567: 333-49.
[PMID: 27232963]

[58] Wang J, Tan Z, Peng J, Qiu Q, Li M. The behaviors of microplastics in the marine environment. Mar Environ Res 2016; 113: 7-17.
[PMID: 26559150]

[59] Prata JC. Microplastics in wastewater: State of the knowledge on sources, fate and solutions. Mar Pollut Bull 2018; 129(1): 262-5.
[PMID: 29680547]

[60] Sun J, Dai X, Wang Q, van Loosdrecht MCM, Ni BJ. Microplastics in wastewater treatment plants: Detection, occurrence and removal. Water Res 2019; 152: 21-37.
[PMID: 30660095]

[61] Zhang Z, Chen Y. Effects of microplastics on wastewater and sewage sludge treatment and their removal: A review. Chem Eng J 2020; 382: 122955.

[62]   Wong JKH, Lee KK, Tang KHD, Yap PS. Microplastics in the freshwater and terrestrial environments: Prevalence, fates, impacts and sustainable solutions. Sci Total Environ 2020; 719: 137512.
[PMID: 32229011]

[63]   Li C, Busquets R, Campos LC. Assessment of microplastics in freshwater systems: A review. Sci Total Environ 2020; 707: 135578.
[PMID: 31784176]

[64]   Li J, Liu H, Paul Chen J. Microplastics in freshwater systems: A review on occurrence, environmental effects, and methods for microplastics detection. Water Res 2018; 137: 362-74.
[PMID: 29580559]

[65]   Woodall LC, Sanchez-Vidal A, Canals M, *et al.* The deep sea is a major sink for microplastic debris. R Soc Open Sci 2014; 1(4): 140317.
[PMID: 26064573]

[66]   Wang F, Wong CS, Chen D, Lu X, Wang F, Zeng EY. Interaction of toxic chemicals with microplastics: A critical review. Water Res 2018; 139: 208-19.
[PMID: 29653356]

[67]   Alimba CG, Faggio C. Microplastics in the marine environment: Current trends in environmental pollution and mechanisms of toxicological profile. Pharma 2019; 68: 61-74.
[PMID: 30877952]

[68]   Choy CA, Robison BH, Gagne TO, *et al.* The vertical distribution and biological transport of marine microplastics across the epipelagic and mesopelagic water column. Sci Rep 2019; 9(1): 7843.
[PMID: 31171833]

[69]   Pohl F, Eggenhuisen JT, Kane IA, Clare MA. Transport and burial of microplastics in deep-marine sediments by turbidity currents. Environ Sci Technol 2020; 54(7): 4180-9.
[PMID: 32142607]

[70]   Saeed T, Al-Jandal N, Al-Mutairi A, Taqi H. Microplastics in Kuwait marine environment: Results of first survey. Mar Pollut Bull 2020; 152: 110880.
[PMID: 31957677]

[71]   Darabi M, Majeed H, Diehl A, Norton J, Zhang Y. A review of microplastics in aquatic sediments: occurrence, fate, transport, and ecological impact. Curr Pollut Rep 2021; (1): 40-53.

[72]   Colton JB Jr, Burns BR, Knapp FD. Plastic particles in surface waters of the northwestern atlantic. Science 1974; 185(4150): 491-7.
[http://dx.doi.org/10.1126/science.185.4150.491] [PMID: 17830390]

[73]   Wong CS, Green DR, Cretney WJ. Quantitative tar and plastic waste distributions in the Pacific Ocean. Nature 1974; 247: 30-2.

[74]   Thompson RC, Olsen Y, Mitchell RP, *et al.* Lost at sea: where is all the plastic? Science 2004; 304(5672): 838.
[http://dx.doi.org/10.1126/science.1094559] [PMID: 15131299]

[75]   Ivar do Sul JA, Costa MF. Marine debris review for Latin America and the wider Caribbean region: from the 1970s until now, and where do we go from here? Mar Pollut Bull 2007; 54(8): 1087-104.
[http://dx.doi.org/10.1016/j.marpolbul.2007.05.004] [PMID: 17624374]

[76]   Eriksson C, Burton H, Fitch S, Schulz M, van den Hoff J. Daily accumulation rates of marine debris on sub-Antarctic island beaches. Mar Pollut Bull 2013; 66(1-2): 199-208.
[http://dx.doi.org/10.1016/j.marpolbul.2012.08.026] [PMID: 23219394]

[77]   Ivar do Sul JA, Costa MF, Barletta M, Cysneiros FJA. Pelagic microplastics around an archipelago of the Equatorial Atlantic. Mar Pollut Bull 2013; 75(1-2): 305-9.
[http://dx.doi.org/10.1016/j.marpolbul.2013.07.040] [PMID: 23953893]

[78]   Enders K, Lenz R, Stedmon CA, Nielsen TG. Abundance, size and polymer composition of marine microplastics ≥10μm in the Atlantic Ocean and their modelled vertical distribution. Mar Pollut Bull 2015; 100(1): 70-81.
[http://dx.doi.org/10.1016/j.marpolbul.2015.09.027]

[79]   Kanhai DK, Officer R, Lyashevska O, Thompson RC, O'Connor I. Microplastic abundance, distribution and composition along a latitudinal gradient in the Atlantic Ocean. Mar Pollut Bull 2017; 115(1-2): 307-14.
[http://dx.doi.org/10.1016/j.marpolbul.2016.12.025] [PMID: 28007381]

[80]   Mendoza LMR, Jones PR. Characterisation of microplastics and toxic chemicals extracted from microplastic samples from the north Pacific gyre. Environ Chem 2015; 12: 611-7.

[81]   Lusher AL, Tirelli V, O'Connor I, Officer R. Microplastics in Arctic polar waters: the first reported values of particles in surface and sub-surface samples. Sci Rep 2015; 5: 14947.
[PMID: 26446348]

[82]   Kane IA, Clare MA. Dispersion, accumulation, and the ultimate fate of microplastics in deep-marine environments: a review and future directions. Front Earth Sci 2019; 30(7): 80.

[83]   Isobe A, Uchida K, Tokai T, Iwasaki S. East Asian seas: A hot spot of pelagic microplastics. Mar Pollut Bull 2015; 101(2): 618-23.
[http://dx.doi.org/10.1016/j.marpolbul.2015.10.042] [PMID: 26522164]

[84]   Cózar A, Echevarría F, González-Gordillo JI, *et al.* Plastic debris in the open ocean. Proc Natl Acad Sci USA 2014; 111(28): 10239-44.
[http://dx.doi.org/10.1073/pnas.1314705111] [PMID: 24982135]

[85]   Claessens M, De Meester S, Van Landuyt L, De Clerck K, Janssen CR. Occurrence and distribution of microplastics in marine sediments along the Belgian coast. Mar Pollut Bull 2011; 62(10): 2199-204.
[http://dx.doi.org/10.1016/j.marpolbul.2011.06.030] [PMID: 21802098]

[86]   Ballent A, Purser A, de Jesus Mendes P, Pando S, Thomsen L. Physical transport properties of marine microplastic pollution. Biogeosciences Discuss 2012; 9: 18755-98.
[http://dx.doi.org/10.5194/bgd-9-18755-2012]

[87]   Barnes DKA, Walters A, Gonçalves L. Macroplastics at sea around Antarctica. Mar Environ Res 2010; 70(2): 250-2.
[http://dx.doi.org/10.1016/j.marenvres.2010.05.006] [PMID: 20621773]

[88]   Ivar do Sul JA, Spengler Â, Costa MF. Here, there and everywhere. Small plastic fragments and pellets on beaches of Fernando de Noronha (Equatorial Western Atlantic). Mar Pollut Bull 2009; 58(8): 1236-8.
[http://dx.doi.org/10.1016/j.marpolbul.2009.05.004] [PMID: 19486997]

[89]   Müller A, Österlund H, Marsalek J, Viklander M. The pollution conveyed by urban runoff: A review of sources. Sci Total Environ 2020; 709: 136125.
[http://dx.doi.org/10.1016/j.scitotenv.2019.136125] [PMID: 31905584]

[90]   Lebreton L, Slat B, Ferrari F, *et al.* Evidence that the great pacific garbage patch is rapidly accumulating plastic. Sci Rep 2018; 8(1): 4666.
[http://dx.doi.org/10.1038/s41598-018-22939-w] [PMID: 29568057]

[91]   Ryan PG, Moore CJ, van Franeker JA, Moloney CL. Monitoring the abundance of plastic debris in the marine environment. Philos Trans R Soc Lond B Biol Sci 2009; 364(1526): 1999-2012.
[http://dx.doi.org/10.1098/rstb.2008.0207] [PMID: 19528052]

[92]   Schmidt C, Krauth T, Wagner S. Export of plastic debris by rivers into the Sea. Environ Sci Technol 2017; 51(21): 12246-53.
[http://dx.doi.org/10.1021/acs.est.7b02368] [PMID: 29019247]

[93]   Savoca S, Capillo G, Mancuso M, *et al.* Microplastics occurrence in the Tyrrhenian waters and in the

gastrointestinal tract of two congener species of seabreams. Environ Toxicol Pharmacol 2019; 67: 35-41.
[http://dx.doi.org/10.1016/j.etap.2019.01.011] [PMID: 30711873]

[94]   Strungaru SA, Jijie R, Nicoara M, Plavan G, Faggio C. Micro-(nano) plastics in freshwater ecosystems: Abundance, toxicological impact and quantification methodology. Trends Analyt Chem 2019; 110: 116-28.

[95]   Antunes J, Frias J, Sobral P. Microplastics on the Portuguese coast. Mar Pollut Bull 2018; 131(Pt A): 294-302.
[http://dx.doi.org/10.1016/j.marpolbul.2018.04.025] [PMID: 29886950]

[96]   Besseling E, Redondo-Hasselerharm P, Foekema EM, Koelmans AA. Quantifying ecological risks of aquatic micro- and nanoplastic. Crit Rev Environ Sci Technol 2019; 49: 32-80.

[97]   Cheung PK, Fok L, Hung PL, Cheung LTO. Spatio-temporal comparison of neustonic microplastic density in Hong Kong waters under the influence of the Pearl River Estuary. Sci Total Environ 2018; 628-629: 731-9.
[http://dx.doi.org/10.1016/j.scitotenv.2018.01.338] [PMID: 29454213]

[98]   Moreira FT, Prantoni AL, Martini B, de Abreu MA, Stoiev SB, Turra A. Small-scale temporal and spatial variability in the abundance of plastic pellets on sandy beaches: Methodological considerations for estimating the input of microplastics. Mar Pollut Bull 2016; 102(1): 114-21.
[http://dx.doi.org/10.1016/j.marpolbul.2015.11.051] [PMID: 26677755]

[99]   Lebreton LCM, van der Zwet J, Damsteeg JW, Slat B, Andrady A, Reisser J. River plastic emissions to the world's oceans. Nat Commun 2017; 8: 15611.
[http://dx.doi.org/10.1038/ncomms15611] [PMID: 28589961]

[100]   Wagner M, Lambert S, Eds. Freshwater microplastics: Emerging environmental contaminants?In: The Handbook of Environmental Chemistry. Cham, Switzeraland: Springer 2018; 58: pp. 113-34.

[101]   Song YK, Hong SH, Eo S, *et al.* Horizontal and vertical distribution of microplastics in Korean coastal waters. Environ Sci Technol 2018; 52(21): 12188-97.
[http://dx.doi.org/10.1021/acs.est.8b04032] [PMID: 30295469]

[102]   Tibbetts J, Krause S, Lynch I, Sambrook Smith GH. Abundance, distribution, and drivers of microplastic contamination in urban river environments. Water 2018; 10(11): 1597.

[103]   Bellasi A, Binda G, Pozzi A, Galafassi S, Volta P, Bettinetti R. Microplastic contamination in freshwater environments: A review, focusing on interactions with sediments and benthic organisms. Environments 2020; 7(4): 30.

[104]   Mani T, Hauk A, Walter U, Burkhardt-Holm P. Microplastics profile along the Rhine River. Sci Rep 2015; 5: 17988.
[http://dx.doi.org/10.1038/srep17988] [PMID: 26644346]

[105]   Campanale C, Stock F, Massarelli C, *et al.* Microplastics and their possible sources: The example of Ofanto river in southeast Italy. Environ Pollut 2020; 258: 113284.
[http://dx.doi.org/10.1016/j.envpol.2019.113284] [PMID: 32005487]

[106]   Kapp KJ, Yeatman E. Microplastic hotspots in the snake and lower Columbia rivers: A journey from the greater Yellowstone ecosystem to the Pacific Ocean. Environ Pollut 2018; 241: 1082-90.
[http://dx.doi.org/10.1016/j.envpol.2018.06.033] [PMID: 30029316]

[107]   Ding L, Mao RF, Guo X, Yang X, Zhang Q, Yang C. Microplastics in surface waters and sediments of the Wei River, in the northwest of China. Sci Total Environ 2019; 667: 427-34.
[http://dx.doi.org/10.1016/j.scitotenv.2019.02.332] [PMID: 30833241]

[108]   Hendrickson E, Minor EC, Schreiner K. Microplastic abundance and composition in western Lake Superior as determined *via* microscopy, Pyr-GC/MS, and FTIR. Environ Sci Technol 2018; 52(4): 1787-96.
[http://dx.doi.org/10.1021/acs.est.7b05829] [PMID: 29345465]

[109] Eriksen M, Mason S, Wilson S, *et al*. Microplastic pollution in the surface waters of the Laurentian Great Lakes. Mar Pollut Bull 2013; 77(1-2): 177-82.
[http://dx.doi.org/10.1016/j.marpolbul.2013.10.007] [PMID: 24449922]

[110] Faure F, Corbaz M, Baecher H, de Alencastro L. Pollution due to plastics and microplastics in Lake Geneva and in the Mediterranean sea. Arch Sci 2012; 65: 157-64.

[111] Ballent A, Corcoran PL, Madden O, Helm PA, Longstaffe FJ. Sources and sinks of microplastics in Canadian Lake Ontario nearshore, tributary and beach sediments. Mar Pollut Bull 2016; 110(1): 383-95.
[http://dx.doi.org/10.1016/j.marpolbul.2016.06.037] [PMID: 27342902]

[112] Migwi FK, Ogunah JA, Kiratu JM. Occurrence and spatial distribution of microplastics in the surface waters of Lake Naivasha, Kenya. Environ Toxicol Chem 2020; 39(4): 765-74.
[http://dx.doi.org/10.1002/etc.4677] [PMID: 32004390]

[113] Hoellein TJ, Shogren AJ, Tank JL, Risteca P, Kelly JJ. Microplastic deposition velocity in streams follows patterns for naturally occurring allochthonous particles. Sci Rep 2019; 9(1): 3740.
[http://dx.doi.org/10.1038/s41598-019-40126-3] [PMID: 30842497]

[114] Kataoka T, Nihei Y, Kudou K, Hinata H. Assessment of the sources and inflow processes of microplastics in the river environments of Japan. Environ Pollut 2019; 244: 958-65.
[http://dx.doi.org/10.1016/j.envpol.2018.10.111] [PMID: 30469290]

[115] Hurley R, Woodward J, Rothwell JJ. Microplastic contamination of river beds significantly reduced by catchment-wide flooding. Nat Geo 2018; (4): 251-7.

[116] Hoellein TJ, Shogren AJ, Tank JL, Risteca P, Kelly JJ. Microplastic deposition velocity in streams follows patterns for naturally occurring allochthonous particles. Sci Rep 2019; 9(1): 3740.
[http://dx.doi.org/10.1038/s41598-019-40126-31] [PMID: 30842497]

[117] Manzoor S, Kaur H, Singh R. Existence of microplastic as pollutant in harike wetland: an analysis of plastic composition and first report on ramsar wetland of india.

[118] KAUR H, SINGH R. Analysis Of Nylon 6 As Microplastic In Harike Wetland By Comparing Its IR Spectra With Virgin Nylon 6 And 6.6. Eur J Mol Clin Med 2020; 7(07): 2020.

[119] Güven O, Gökdağ K, Jovanović B, Kıdeyş AE. Microplastic litter composition of the Turkish territorial waters of the Mediterranean Sea, and its occurrence in the gastrointestinal tract of fish. Environ Pollut 2017; 223: 286-94.
[http://dx.doi.org/10.1016/j.envpol.2017.01.025] [PMID: 28117186]

[120] Rillig MC. Microplastic in terrestrial ecosystems and the soil? Environ Sci Technol 2012; 46(12): 6453-4.
[http://dx.doi.org/10.1021/es302011r] [PMID: 22676039]

[121] McCormick A, Hoellein TJ, Mason SA, Schluep J, Kelly JJ. Microplastic is an abundant and distinct microbial habitat in an urban river. Environ Sci Technol 2014; 48(20): 11863-71.
[http://dx.doi.org/10.1021/es503610r] [PMID: 25230146]

[122] Xu B, Liu F, Cryder Z, *et al*. Microplastics in the soil environment: occurrence, risks, interactions and fate–a review. Crit Rev Environ Sci Technol 2020; 50: 2175-222.
[http://dx.doi.org/10.1080/10643389.2019.1694822]

[123] Guo JJ, Huang XP, Xiang L, *et al*. Source, migration and toxicology of microplastics in soil. Environ Int 2020; 137: 105263.
[http://dx.doi.org/10.1016/j.envint.2019.105263] [PMID: 32087481]

[124] Hurley RR, Nizzetto L. Fate and occurrence of micro (nano) plastics in soils: Knowledge gaps and possible risks. Curr Opin Environ Sci Health 2018; 1: 6-11.

[125] Alimi OS, Farner Budarz J, Hernandez LM, Tufenkji N. Microplastics and nanoplastics in aquatic environments: aggregation, deposition, and enhanced contaminant transport. Environ Sci Technol

2018; 52(4): 1704-24.
[http://dx.doi.org/10.1021/acs.est.7b05559] [PMID: 29265806]

[126] Qi R, Jones DL, Li Z, Liu Q, Yan C. Behavior of microplastics and plastic film residues in the soil environment: A critical review. Sci Total Environ 2020; 703: 134722.
[http://dx.doi.org/10.1016/j.scitotenv.2019.134722] [PMID: 31767311]

[127] Li L, Geng S, Wu C, *et al.* Microplastics contamination in different trophic state lakes along the middle and lower reaches of Yangtze River Basin. Environ Pollut 2019; 254(Pt A): 112951.
[http://dx.doi.org/10.1016/j.envpol.2019.07.119] [PMID: 31374488]

[128] Liu K, Wang X, Wei N, Song Z, Li D. Accurate quantification and transport estimation of suspended atmospheric microplastics in megacities: Implications for human health. Environ Int 2019; 132: 105127.
[http://dx.doi.org/10.1016/j.envint.2019.105127] [PMID: 31487610]

[129] Dris R, Gasperi J, Saad M, Mirande C, Tassin B. Synthetic fibers in atmospheric fallout: A source of microplastics in the environment? Mar Pollut Bull 2016; 104(1-2): 290-3.
[http://dx.doi.org/10.1016/j.marpolbul.2016.01.006] [PMID: 26787549]

[130] Nizzetto L, Futter M, Langaas S. Are agricultural soils dumps for microplastics of urban origin? Environ Sci Technol 2016; 50(20): 10777-9.
[http://dx.doi.org/10.1021/acs.est.6b04140] [PMID: 27682621]

[131] Bläsing M, Amelung W. Plastics in soil: Analytical methods and possible sources. Sci Total Environ 2018; 612: 422-35.
[http://dx.doi.org/10.1016/j.scitotenv.2017.08.086] [PMID: 28863373]

[132] Ruíz-Machuca LM, Ibarra-Jiménez L, Valdez-Aguilar LA, Robledo-Torres V, Benavides Mendoza A, Cabrera-De La Fuente M. Cultivation of potato–use of plastic mulch and row covers on soil temperature, growth, nutrient status, and yield. Acta Agric Scand B Soil Plant Sci 2015; 65: 30-5.

[133] Hebner TS, Maurer-Jones MA. Characterizing microplastic size and morphology of photodegraded polymers placed in simulated moving water conditions. Environ Sci Process Impacts 2020; 22(2): 398-407.
[http://dx.doi.org/10.1039/c9em00475k] [PMID: 31993606]

[134] Petersen F, Hubbart JA. The occurrence and transport of microplastics: The state of the science. Sci Total Environ 2021; 758: 143936.
[http://dx.doi.org/10.1016/j.scitotenv.2020.143936] [PMID: 33333307]

[135] Steinmetz Z, Wollmann C, Schaefer M, *et al.* Plastic mulching in agriculture. Trading short-term agronomic benefits for long-term soil degradation? Sci Total Environ 2016; 550: 690-705.
[http://dx.doi.org/10.1016/j.scitotenv.2016.01.153] [PMID: 26849333]

[136] Carr SA, Liu J, Tesoro AG. Transport and fate of microplastic particles in wastewater treatment plants. Water Res 2016; 91: 174-82.
[http://dx.doi.org/10.1016/j.watres.2016.01.002] [PMID: 26795302]

[137] He P, Chen L, Shao L, Zhang H, Lü F. Municipal solid waste (MSW) landfill: A source of microplastics? -Evidence of microplastics in landfill leachate. Water Res 2019; 159: 38-45.
[http://dx.doi.org/10.1016/j.watres.2019.04.060] [PMID: 31078750]

[138] O'Connor D, Pan S, Shen Z, *et al.* Microplastics undergo accelerated vertical migration in sand soil due to small size and wet-dry cycles. Environ Pollut 2019; 249: 527-34.
[http://dx.doi.org/10.1016/j.envpol.2019.03.092] [PMID: 30928524]

[139] Cao D, Wang X, Luo X, Liu G. Zheng. Effects of polystyrene microplastics on the fitness of earthworms in an agricultural soil. IOP C Ser. Earth Env 2017; 61: 012148.
[http://dx.doi.org/10.1088/1755-1315/61/1/012148]

[140] Rillig MC, Ziersch L, Hempel S. Microplastic transport in soil by earthworms. Sci Rep 2017; 7(1): 1362.

[http://dx.doi.org/10.1038/s41598-017-01594-7] [PMID: 28465618]

[141]   Zhang S, Wang J, Yan P, *et al.* Non-biodegradable microplastics in soils: a brief review and challenge. J Hazar. Mat 2021; 409: 124-525.

[142]   Zhu F, Zhu C, Wang C, Gu C. Occurrence and ecological impacts of microplastics in soil systems: a review. Bull Environ Contam Toxicol 2019; 102(6): 741-9.
[http://dx.doi.org/10.1007/s00128-019-02623-z] [PMID: 31069405]

[143]   Yousuf A, Naseer M, Naqash N, Singh R. Isolation and identification of microplastic particles from agricultural soil and its detection by fluorescence microscope technique. Think India Journal 2019; 22(16): 3934-49.

[144]   Manzoor S, Naqash N, Rashid G, Singh R. Plastic material degradation and formation of microplastic in the environment: A review. Mater Today Proc 2021.

[145]   Wang W, Ge J, Yu X, Li H. Environmental fate and impacts of microplastics in soil ecosystems: Progress and perspective. Sci Total Environ 2020; 708: 134841.
[http://dx.doi.org/10.1016/ j.scitotenv.2019.134841] [PMID: 31791759]

[146]   Dris R, Gasperi J, Rocher V, Saad M, Renault N, Tassin B. Microplastic contamination in an urban area: a case study in Greater Paris. Environ Chem 2015; 12(5): 592-9.
[http://dx.doi.org/10.1071/EN14167]

[147]   Enyoh CE, Verla AW, Verla EN, Ibe FC, Amaobi CE. Airborne microplastics: a review study on method for analysis, occurrence, movement and risks. Environ Monit Assess 2019; 191(11): 668.
[http://dx.doi.org/10.1007/s10661-019- 7842-0] [PMID: 31650348]

[148]   Enyoh CE, Verla AW, Verla EN, Ibe FC, Amaobi CE. Airborne microplastics: a review study on method for analysis, occurrence, movement and risks. SN Applied Sciences 2019.

[149]   Yumei H, Xian Q, Wenjing W, Gong H, Jun W. Mini-review on current studies of airborne microplastics: Analytical methods, occurrence, sources, fate and potential risk to human beings. Trends in Analytical Chemistry 2020; 125.

[150]   Klein M, Fischer EK. Microplastic abundance in atmospheric deposition within the Metropolitan area of Hamburg, Germany. Sci Total Environ 2019; 685: 96-103.
[http://dx.doi.org/10.1016/j.scitotenv.2019.05.405] [PMID: 31174127]

[151]   Cai L, Wang J, Peng J, *et al.* Characteristic of microplastics in the atmospheric fallout from Dongguan city, China: preliminary research and first evidence. Environ Sci Pollut Res 2017; 24: 24928-35.
[http://dx.doi.org/10.1007/s11356-017-0116-x]

[152]   Bank M S, Hansson S V. The plastic cycle: a novel and holistic paradigm for the anthropocene. Environ Sci Technol. ACS 2019; 53: pp. (13)7177-9.
[http://dx.doi.org/10.1021/acs.est.9b02942]

[153]   Evangeliou N, Grythe H, Klimont Z, *et al.* Atmospheric transport is a major pathway of microplastics to remote regions. Nat Commun 2020; 11(1): 3381.
[http://dx.doi.org/10.1038/s41467-020-17201-9] [PMID: 32665541]

[154]   Mbachu O, Jenkins G, Pratt C. Kaparaju, P. A new contaminant superhighway? a review of sources, measurement techniques and fate of atmospheric microplastics. Water Air Soil Pollut. Springer 2020; 231 .(2).
[http://dx.doi.org/10.1007/s11270-020-4459-4]

[155]   Dris R, Gasperi J, Mirande C, *et al.* A first overview of textile fibers, including microplastics, in indoor and outdoor environments. Environ Pollut 2017; 221: 453-8.
[http://dx.doi.org/10.1016/j.envpol. 2016.12.013.]

[156]   Beaurepaire M, Dris R, Gasperi J, Tassin B. Microplastics in the atmospheric compartment: a comprehensive review on methods, results on their occurrence and determining factors. Current Opinion in Food Science, Elsevier 2021; 41: 16.

[157] Allen S, Allen D, Phoenix VR, *et al.* Atmospheric transport and deposition of microplastics in a remote mountain catchment. Nat Geosci 2019; 12(5): 339-44.

[158] Bergmann M, Mützel S, Primpke S, Tekman MB, Trachsel J, Gerdts G. White and wonderful? Microplastics prevail in snow from the Alps to the Arctic. Sci Adv 2019; 5(8): eaax1157.
[http://dx.doi.org/10.1126/sciadv.aax1157] [PMID: 31453336]

[159] Rezaei M, Riksen MJPM, Sirjani E, Sameni A, Geissen V. Wind erosion as a driver for transport of light density microplastics. Sci Total Environ 2019; 669: 273-81.
[http://dx.doi.org/10.1016/j.scitotenv.2019.02.382] [PMID: 30878934]

[160] Genc ANM, Vural N, Balas L. Modeling transport of microplastics in enclosed coastal waters: A case study in the Fethiye Inner Bay. Mar Pollut Bull 2020; 150: 110747.
[http://dx.doi.org/10.1016/j.marpolbul.2019.110747] [PMID: 31784264]

[161] Sommer F, Dietze V, Baum A, *et al.* Tire abrasion as a major source of microplastics in the environment. Aerosol Air Qual Res 2018; 18(8): 2014-28.

[162] Allen S, Allen D, Moss K, Le Roux G, Phoenix VR, Sonke JE. Examination of the ocean as a source for atmospheric microplastics. PLoS One 2020; 15(5): e0232746.
[http://dx.doi.org/10.1371/journal.pone.0232746] [PMID: 32396561]

[163] Dehghani S, Moore F, Akhbarizadeh R. Microplastic pollution in deposited urban dust, Tehran metropolis, Iran. Environ Sci Pollut Res Int 2017; 24(25): 20360-71.
[http://dx.doi.org/10.1007/s11356-017-9674-1] [PMID: 28707239]

[164] Wang W, Gao H, Jin S, Li R, Na G. The ecotoxicological effects of microplastics on aquatic food web, from primary producer to human: A review. Ecotoxicol Environ Saf 2019; 173: 110-7.
[http://dx.doi.org/10.1016/j.ecoenv.2019.01.113] [PMID: 30771654]

[165] Zhu D, Chen QL, An XL, *et al.* Exposure of soil collembolans to microplastics perturbs their gut microbiota and alters their isotopic composition. Soil Biol Biochem 2018; 116: 302-10.

[166] Sussarellu R, Suquet M, Thomas Y, *et al.* Oyster reproduction is affected by exposure to polystyrene microplastics. Proc Natl Acad Sci USA 2016; 113(9): 2430-5.
[http://dx.doi.org/10.1073/pnas.1519019113] [PMID: 26831072]

[167] Rochman CM, Kurobe T, Flores I, Teh SJ. Early warning signs of endocrine disruption in adult fish from the ingestion of polyethylene with and without sorbed chemical pollutants from the marine environment. Sci Total Environ 2014; 493: 656-61.
[http://dx.doi.org/10.1016/j.scitotenv.2014.06.051] [PMID: 24995635]

[168] Rainieri S, Conlledo N, Larsen BK, Granby K, Barranco A. Combined effects of microplastics and chemical contaminants on the organ toxicity of zebrafish (*Danio rerio*). Environ Res 2018; 162: 135-43.
[http://dx.doi.org/10.1016/j.envres.2017.12.019] [PMID: 29306661]

[169] Mercogliano R, Avio C G, Regoli F, Anastasio A, Colavita G, Santonicola S. Occurrence of microplastics in commercial seafood under the perspective of the human food chain. A review J Agri Food Chem 2020; 20,68(19): 5296-301.

[170] Rochman CM, Brookson C, Bikker J, *et al.* Rethinking microplastics as a diverse contaminant suite. Environ Toxicol Chem 2019; 38(4): 703-11.
[http://dx.doi.org/10.1002/etc.4371] [PMID: 30909321]

[171] Qi Y, Yang X, Pelaez AM, *et al.* Macro- and micro- plastics in soil-plant system: Effects of plastic mulch film residues on wheat (*Triticum aestivum*) growth. Sci Total Environ 2018; 645: 1048-56.
[http://dx.doi.org/10.1016/j.scitotenv.2018.07.229] [PMID: 30248830]

[172] Huerta Lwanga E, Gertsen H, Gooren H, *et al.* Incorporation of microplastics from litter into burrows of *Lumbricus terrestris.* Environ Pollut 2017; 220(Pt A): 523-31.
[http://dx.doi.org/10.1016/j.envpol.2016.09.096] [PMID: 27726978]

[173] Naqash N, Manzoor S, Singh R. Microplastic hazard, management, remediation, and control strategies: a review. Int J Environ Technol Manag 2022; 1(1): 1.
[http://dx.doi.org/10.1504/IJETM.2022.10049175]

[174] Anbumani S, Kakkar P. Ecotoxicological effects of microplastics on biota: a review. Environ Sci Pollut Res Int 2018; 25(15): 14373-96.
[http://dx.doi.org/10.1007/s11356-018-1999-x] [PMID: 29680884]

[175] Karbalaei S, Hanachi P, Walker TR, Cole M. Occurrence, sources, human health impacts and mitigation of microplastic pollution. Environ Sci Pollut Res Int 2018; 25(36): 36046-63.
[http://dx.doi.org/10.1007/s11356-018-3508-7] [PMID: 30382517]

[176] Gasperi J, Wright SL, Dris R, Collard F, Mandin Guerrouache M, Langlois Kelly FJ. Tassim. Microplastics in air- we are breathing in it. Curr Opin Environ Sci Health 2018; I: 1-5.

[177] Mohsen P, Eric L, Didier R, Chuanyi W. Review of microplastics from environment - A review. Environ Chem Lett 2020; 18: 807-28.
[http://dx.doi.org/10.1007/s10311-020-00983-1]

[178] He P, Chen L, Shao L, Zhang H, Lü F. Municipal solid waste (MSW) landfill: A source of microplastics? -Evidence of microplastics in landfill leachate. Water Res 2019; 159: 38-45.
[http://dx.doi.org/10.1016/j.watres.2019.04.060] [PMID: 31078750]

[179] Marine plastic litter and microplastics. United Nations Environment Programme 2016.

[180] Austin HP, Allen MD, Donohoe BS, *et al.* Characterization and engineering of a plastic-degrading aromatic polyesterase. Proc Natl Acad Sci USA 2018; 115(19): E4350-7.
[http://dx.doi.org/10.1073/pnas.1718804115] [PMID: 29666242]

[181] Adam I, Walker TR, Bezerra JC, Clayton A. Policies to reduce single-use plastic marine pollution in West Africa. Mar Policy 2020; 116: 103-928.

[182] Bezerra JC, Walker TR, Clayton CA, Adam I. Single-use plastic bag policies in the Southern African development community. Environ Chall 2021; 3: 1000-29.

CHAPTER 3

Impact of Microplastics on Flora and Fauna

Quseen Mushtaq Reshi[1], **Imtiaz Ahmed**[1,*], **Ishtiyaq Ahmad**[1] and **Francesco Fazio**[2]

[1] *Fish Nutrition Research Laboratory, Department of Zoology, University of Kashmir, Srinagar-190006, Jammu and Kashmir, India*

[2] *Department of Veterinary Science, University of Messina, Messina, Italy*

Abstract: Microplastics are the compound class of greatly altered, synthetic particulates, which pollute wide-ranging types of environments. Being an impending source of concern owing to wide variability in their size range makes them potentially dangerous at all trophic levels. Numerous studies have studied the harmful effects of microplastics on the biota. The present study aims to compile information about the effect of microplastics on various species belonging to different taxonomic groups as reported from different parts of the world based on which, a general overview has been generated which clearly emphasizes that substantial efforts are required to deeply investigate the abundance, distribution and effects of microplastics on the flora and fauna of both terrestrial as well as aquatic ecosystems. However, the influence of microplastic contamination on human health and plants has received less intention. The knowledge derived from various studies clearly indicates that in order to safeguard our environment from the deteriorating effects of microplastics, we need to thoroughly control the amount of plastic production. Moreover, stress should be laid to make more use of bio-degradable products so as to minimize the demand for these plastic materials. Also, there is a dire need to aware the masses about the harmful effects of microplastics and the adoption of such policies at the global level which formulate a strong action plan for solid waste management so as to alleviate microplastic pollution, which otherwise could threaten ecological balance as well as harm the health and survival of various species.

Keywords: Aquatic organisms, Contamination, Human health, Microplastics, Plants, Pollution.

INTRODUCTION

The earth is persistently facing numerous threats by virtue of natural disasters and pollution caused due to anthropogenic activities [1]. Almost all natural resources are badly affected due to the frequent changes present in the environment.

---

* **Corresponding Author Imtiaz Ahmed:** Fish Nutrition Research Laboratory, Department of Zoology, University of Kashmir, Srinagar-190006, Jammu and Kashmir, India E-mail: imtiazamu1@yahoo.com

However, the aquatic ecosystem gets more affected due to all these changes in the form of water pollution [2, 3]. Plastic and microplastic pollution, along with associated nanoparticles, is one of the 'emerging' concerns in the mind of scientists at the moment [4]. Among them, microplastics are presently considered a hidden/silent agent that badly affects many aquatic organisms, therebaffectinged the food chain/web. Microplastics are defined as microscopic-sized plastic materials with a diameter of less than 5mm, which largely arise from parts of larger plastics directly released into the environment as production of personal care, UV radiation and mechanical degradation of plastic bags, bottles, food wrappers, *etc* [5]. Microplastics represent an obstinate environmental issue because of their existence in aquatic as well as terrestrial ecosystems, and the difficulty in detecting these items [6, 7]. The microplastics are actually made of a multifarious congregation of units having different sizes, shapes, densities, colours and chemical entities, which could have significance on their transportation methods and subsequent destinies in aquatic surroundings [8]. So, the question is how microplastics are formed in the aquatic environment. The answer would be it could be due to the direct source in the form of waste release from industries or an indirect source like wind and rain that cause deposition of piled wastes from houses that eventually end up into the water bodies and finally enter into the oceans [9]. Microplastics are sub-divided as primary and secondary forms. Primary microplastics refer to such materials which arise directly because of the usage of such products or materials which are made up of microplastics as their components. Secondary microplastics are those which are made as a result of the disintegration of plastics of larger sizes [6, 10]. Infiltration of microplastics into multiple ecosystems elevates concerns of their potential toxicity thereby causing harm to both flora and fauna. SAPEA (2019) report indicated the qualitative information and summarised the presence of microplastics in wastewater, drinking water, freshwater, estuarine, coastal and marine environments, besides air, soils, biota as well as food [11]. However, it further suggested that there is a need to understand the precise quantitative presence of microplastics in the atmosphere, soil and the marine water column. They also stressed upon the ambiguity or non-existence of data on microplastics less than 300 μm and limitations in comparability of data, which inhibits the understanding of the microplastic's impact on the environment.

Nowadays, the possible chances of cascading of harmful effects of microplastics through the trophic layers of the ecosystem have been one of the main concerns of analysts. Numerous studies on calcification and ecotoxicity-based influences of microplastics on organisms have indicated that the ingestion of these plastics gravely affects the animals [12]. Some of the common issues include obstruction of gut, loss of lipid stores, elevated immune response, disorder in other normal physiological processes related to respiration, photosynthesis, reproduction, *etc.*

FPA-based Micro-FTIR (Focal Plane Array-based Micro-Foutier Transform Infrared) and Micro-Raman Spectroscopy are the analytical laboratory techniques used to measure the presence of microplastics [13]. Although durability and longevity are among the most valued merits of plastic products but these plastics when not managed well, often lead to contamination of the surrounding in both freshwater as well as marine environments. Also, products made of plastic degrade very slowly especially on exposure to ultraviolet radiation emitted by the sun and high temperatures. However, this degradation mainly causes the breakdown of the particles into very small sizes fragmented from macroscopic size to microscopic size and finally to indiscernible measurements called nanoplastics [14].

Microplastic presence in the flora and fauna largely affected the whole system in many ways and some are discussed below:

INFLUENCE OF MICROPLASTICS IN TERRESTRIAL ECOSYSTEMS

The ever-increasing usage of plastics as a packaging material and in various other forms has led to a massive accumulation of microplastics in our atmosphere. Microplastics have been detected in atmospheric fallout indicating their dispersion into and transportation *via* the air [15]. Research work assessing the impact of microplastics on land-dwelling creatures shows the occurrence of microplastics in livestock [16] and birds [17] and the ingested microplastics may discharge constituents which could be potential disrupters of the organism's endocrine system [18]. Analogous to microplastics in the aquatic environment, terrestrial microplastics can be relocated and biomagnified along the different levels of the food chain. High amount of microplastics in the soil has been reported to affect the biophysical properties of soil badly, along with its biota such as earthworms whose life processes like growth and reproduction drop because the microbes of their gut were disturbed by ingestion of these microplastics [15, 19].

Although many researchers have worked on the influences of microplastics on fauna, however, lesser work has been carried out on plants and agriculture. It was established that farmlands hold considerable microplastics since slush from sewage treatment and wastewater set up are frequently used as additives for soil in the agriculture [20]. In fact, wastewater as well as sewage treatment setups are substantial sources that contribute to environmental microplastics. The sludge in agricultural purposes often causes localized microplastic contamination which afterwards disseminates through transportation by either runoffs or soil and air biota [21, 22]. Till date there is limited statistical information available on the occurrence of microplastics in the land dwelling biota. However, experimental trials have revealed that the worms do take up these particles as a result of an exposure [23]. Earthworms are also able to transfer microplastics within the soil systems [24]. On the other hand, animals having microplastics in their gastrointestinal tracts (GIT) were reported to have comparatively less feeding and filtering activity resulting in poorer levels of energy and fat reserves [25]. They have also been found to trigger immune response and causing inflammation in the organism [26, 27]. Some organisms have also been reported to exhibit liver damage owing to metabolic stress and accumulation of these pollutants on its surface [28]. Increased mortality due to effect on physical and physiological health was reported [29]. Sussarellu *et al.* reported that these pollutants reduced fecundity and also impaired the quality of gamete [30].

A large number of living organisms are often exposed to microplastics as a result of ingesting plastic substances, either because of misidentifying plastic for food or ingesting plastic with other foodstuff [31]. Ingested plastic material may also be transferred through levels in the food chain, however, it still remains unclear how much concentration rises in the higher levels of the trophic chain [32]. The occurrence of microplastics could also alter the habitat of an organism [31]. For accurate assessments of possible damage, information on uptake, exposures, and effects is required.

INFLUENCE OF MICROPLASTICS ON AQUATIC HABITATS

Most shopping bags are made of plastic and end up either in water bodies or in the landfill where they are sometimes swallowed by animals. These animals subsequently suffer a slow, painful and terrible death, even then, the plastic bag still remains undecomposed causing more damage to the environment. For example, the commonly sold items with drinks *i.e.*, Straws are the most common rubbish items found usually around the beaches. Similar to plastic bags, straws are also often swallowed by marine creatures that subsequently cause blockages and ultimately lead to their death. Box bands and other loops are made of plastic which often entangles land as well as sea animals to the extent that they are strangled to death. EFSA (2016), reported the presence of microplastics in wild-caught species comprising even those creatures which are consumed as seafood (fish and shellfish). The impact of pollution caused by plastic materials in the marine environment is considerably documented, however, comparatively there are much lesser studies on the magnitude of pollution in treated water sources as well as freshwater bodies are presently reported. Microplastics are asserted to affect aquatic fauna physically by causing blocking of gills and digestive tracts, behaviourally and physiologically through disruption of various enzymes and hormones as well [33]. The capability of microplastics to adsorb various chemicals such as antimicrobials, polyaromatic hydrocarbons and antibiotics results in the concentration of unsafe chemicals in the aquatic ecosystems, thereby increasing ecotoxicity [33]. FAO (2017) reported that fewer facts were existing for estuarine and freshwater environments, nonetheless, marine environments are highly contaminated with microplastics on a global scale [34]. The genera associated to aquaculture and fisheries were particularly emphasised, but the report asserted that still there was neither direct proof of transfer of microplastics in the wild species through trophic levels nor any evidence based on field studies that could show that ingestion of microplastics affected populations or communities.

The density of microplastics is a determinant factor of their vertical location with respect to the water column, which consecutively could probably affect the probability of fish habitat in different water zones to come across these microplastics [35]. Microplastics in the form of fragments as well as fiber represent the frequently identified forms in different fish species, which are in agreement with their preponderance in waters throughout the globe [36]. After ingesting, most of the microplastics are unfortunately retained in the fish's gastrointestinal tract [37]. Moreover, microplastics may possibly stick to the fish skin or at times to other tissues of the body such as the liver, gills and muscle [38, 39].

Thus the occurrence of plastic pollutants in the environment contaminates the pristine environments, moreover, the direct settling of the particles into the soil and water resources drastically impacts the biota. In the present review, we tried to represent the effect of microplastics on various organisms on the basis of literature and the data are presented in Table **1**.

Freshwater Biota

Freshwater systems act as vectors for the transfer of microplastics with respect to land as well as seawater systems, and they also serve as the sink areas for microplastics. Biota inhabiting all the compartments of freshwater (water column, sediment) react with microplastics, even though the available data is inadequate as compared to the marine environment. The most studied form of interaction through all the sections is primarily ingestion. Fish stands as one of the most widely investigated taxonomic groups from the freshwater environment in relation to ingestion of microplastic. The presence of microplastics and its possible effects on freshwater fish species have been studied such as Nile tilapia, *Oreochromis niloticus*; Nile perch, *Lates niloticus* [83]; common roach, *Rutilus rutilus* [83]; goby, *Neogobius melanostomus* and barbell *Barbus* spp [84].

Table 1. Showing the impact of Microplastics on various taxonomic groups.

Group	Species	Concentration	Sample	Effects	Reference
Fish	*Hyporhamphus intermedius*	(3.7 ± 2.2 items/individual)	saltwater pelagic fish	NA	[40]
	Pomatoschistus microps	100 items/L	NA	Decrease in predatory efficiency and performance of fish	[41]
	Coilia ectenes	(4.0 ± 1.8 items/individual)	saltwater pelagic fish	NA	[40]
	Carassius carassius	15–76 items/fish	NA	The weight of the experimental fish declined by nearly 17.5–21.5% by microplastic exposure.	[42]
	Harpodon neherus	(3.8 ± 2.0 items/individual)	saltwater benthopelagic fish	NA	[40]
	Larimitchys crocea	(4.6 ± 3.4 items/individual)	Saltwater benthopelagic fish	NA	[40]
	Clarias gariepinus	50–500 µg/L	NA	High-density lipoprotein values in the blood of fish were considerably decreased. The ratio of Plasma albumin to globulin increased. Significant decrease in transcription levels of tryptophan hydroxylase.	[43]
	Cyoglossus abreviatus	(6.9 ± 2.4 items/individual)	saltwater demersal fish	NA	[40]
	Thamnaconus septrentionalis	(7.2 ± 2.8 items/individual)	saltwater demersal fish	NA	[40]
	Callionymus planus	(4.8 ± 2.3 items/individual)	saltwater benthic fish,	NA	[40]
	Pseudorasbora parva	(2.5 ± 1.8 items/individual)	Freshwater benthopelagic fish	NA	[40]
	Hypophthalmichthys molitrix	(3.8 ± 2.0 items/individual)	Freshwater benthopelagic fish	NA	[40]
	common goby, *Pomatoschitus microps*	NA	NA	Caused impairment in the aerobic pathway of energy production.	[44]
	Mediterranean mussel (*M. galloprovincialis*	NA	NA	Neurotoxicity, affects the antioxidant system, proliferation of peroxisomes *etc.*	[45]
	Zebrafish (*Danio rerio*)	NA	NA	Intestinal inflammation, intestinal toxicity, intestinal permeability and oxidative stress.	[46]
	Danio rerio	NA	NA	Bioaccumulation in the head and viscera, inhibition of acetylcholinesterase, and the central nervous system showed up-regulation of myelin basic protein.	[47]
	Alepisaurus ferox	NA	mesopelagic fish	NA	[48]
	herring,, Atlantic mackerel, horse mackerel, haddock cod,	NA	NA	NA	[49]
	Japanese medaka (*Oryzias latipes*)	NA	NA	NA	[28]
	Jacopever (*Sebastes schlegelii*	NA	NA	Weight loss Decrease in specific growth and gross energy value.	[50]
	Zebrafish (*Danio rerio*)	NA	NA	Microplastics decrease the bioavailability of phenanthrene and 17α-ethinylestradiol.	[51]
	Common carp (*Cyprinus carpio*)	NA	NA	Increase in the concentration of creatinine and glucose while decrease of total protein, globulins, triglycerides and cholesterol.	[52]
	European seabass (*Dicentrarchus labrax*)	NA	NA	Decrease in swimming velocity, lethargic and erratic swimming Behaviour.	[53]
	Red tilapia (*Oreochromis niloticus*)			Amplified the bioaccumulation of roxithromycin in tissues of the fish.	[54]

(Table 1) cont.....

Group	Species	Concentration	Sample	Effects	Reference
Arthropoda	*D. magna*	NA	NA	Physical damage.	[45]
	Paracyclopina nana	NA	NA	Increase of reactive oxygen species.	[55]
	Calanus helgolandicus	NA	NA	Reduced reproductive potential because of a decrease in energy.	[56]
	Chinese mitten crab (*Eriocheir sinensis*	NA	NA	Influenced intestinal microflora and immune enzyme activity.	[57]
	Norway lobster, *Nephrops norvegicus*	NA	NA	Withholding in spite of excretion through ecdysis, because of complex gut structure	[58]
	common littoral crab, *Carcinus maenas*	NA	NA	Diminished consumption of food and also energy budgets available for growth.	[59]
	herring, *Clupea harengus*	NA	NA	NA	[49]
	horse mackerel, *Trachurus trachurus*	NA	NA	NA	[60]
	Norway lobster, *N. norvegicus*	NA	NA	NA	[61]
	Gooseneck barnacle, *Lepas* spp.)	NA	NA	NA	[62]
	freshwater flea, *Daphnia pulex,*	NA	NA	Decreased growth rate, body length and relative gene expression of heat shock proteins (*HSP70, HSP90*) and antioxidants.	[57]
Annelida	*Arenicola marina*	NA	NA	Increased mortality rate, oxidative stress	[56]
	Arenicola marina	NA	NA	Energy depletion, decreased feeding activity and gut inflammation.	[25]
Chordata	*Homo sapiens*	NA	NA	Minimum of 78,000 particles through tap water intake and nearly 39,000 particles through food (USA).	[63]
		NA	NA	Toxicity through dermal exposure.	[64]
		NA	NA	High cellular toxicity.	[65]
		NA	NA	Neurological effects, cancer and endocrine disruption.	[66, 67]
		NA	NA	Autophagic degradation in endoplasmic reticulum of lung epithelial cells.	[68]
		NA	NA	Alteration in mitochondrial and metabolic activities of placental trophoblast cells.	[69]
		NA	NA	Disrupt the mitochondrial membrane potential.	[70]
	Sea turtle	NA	NA	NA	[71]
-	northern fulmars (*Fulmarus glacialis*),	NA	NA	NA	[72]
-	flesh-footed shearwater (*Puffinus carneipes*)	NA	NA	NA	[73]
-	Laysan albatross (*Phoebastria immutabilis*)	NA	NA	Dehydration, deteriorates nutrition.	[74]
-	tube-nosed seabirds	NA	NA	Decreases food capacity of stomach.	[75]
-	Baleen whale, *Megaptera novaeangliae*	NA	NA	NA	[76]
Bivalves	*Crassostrea gigas* and *Mytilus edulis*	0.47 ± 0.16 particles/g of tissue	NA	NA	[77]
	Alectryonella plicatula, Cyclina sinensis, Meretrix lusoria, Mytilus galloprovincialis;Scapharca subcrenata, Tegillarca granosa; Yesso, *Patinopecten yessoensis, Ruditapes philippinarum, Sinonovacula constricta*	NA	NA	NA	[78, 79, 80]
	Blue mussel, *Mytilus edulis*	NA		Advanced destabilization and accumulation of lysosomal, along with an inflammatory response.	[27]
	Blue mussel	NA	NA	Alteration in the behaviour indicated by making pseudofaeces besides reduced filtering activity.	[24]

(Table 1) cont.....

Group	Species	Concentration	Sample	Effects	Reference
Nematodes	*Caenorhabditis elegans*	NA	NA	Reduced movement behavior, stimulation of reactive oxygen species in the intestines and the nuclear translocation of protein-encoding genes.	[81]
	Caenorhabditis elegans	NA	NA	Decreased brood size, effect on locomotion behavior, augmented permeability of intestines, reactive oxygen species production, amplified the expression of antioxidant defense system and mitochondrial respiratory chain.	[82]

While some other taxonomic groups include bivalves, *Corbicula fluminea* [39] and oligochaetes, *Tubifex* spp [85]. were also studied. 97 percent of macroinvertebrates examined at Tejo estuary had no less than 1 microfibre per individual. Also, around 74 percent of the shorebird faeces studied from the same area did contain microfibres. This clearly hints towards the transfer of microplastics from one trophic level to another. However, the workers stated that the variance in microfibre concentrations observed in shorebirds was established on individuals and not the foraging strategies [86]. In aquaculture, a similar result might take place if microplastics are inadvertently accumulated in the feed during the production process because of already contaminated individuals [34] such as small pelagic fishes which are commonly used in fishmeal and for fish oil because these fishes are at higher risk of mistaking microplastics as food and directly ingesting it. Numerous recent publications have presented information on the basis of the occurrence of microplastics in freshwater surroundings [87, 88, 89, 90]. There is huge spatial unevenness in the distribution pattern of microplastics particularly in freshwater environments, with outcomes presenting manifold differences among and within the studies. Overall, sediments depict higher concentrations as compared to water samples, and outcomes are similar to those achieved from the habitats of coastal marine areas. However, in the case of lacustrine atmospheres, microplastics exhibit higher abundance if close to industrial and urban centres, on the other hand, in some riverine systems such type of relationship does not exist and this could be because of the dynamics of river flow as well as flooding nature of the river [91]. Moreover, windblown microplastics and poor waste management may also add to pollution in comparatively isolated freshwater environments [92].

Marine Biota

The ingestion of microplastics by marine biota has been well documented in some marine species, however, awareness of the occurrence of microplastics in marine biota is still inadequate. Microplastics enter into aquatic environments *via* diverse pathways and these pollutants have been reported from all ecological backgrounds (surface waters, sediments, beaches and water columns). During the manufacturing process, various chemicals called additives are mixed with the microplastics that proficiently sorb persistent, bioaccumulative and toxic contaminants (PBTs). The microplastics consumption by organisms and the build-up of PBTs have been professed as chief hazards causing great risk to marine organisms [34]. The presence of plastic and microplastic pollutants has been detected in mussels and cod inhabiting the coast of Norwegian [93, 94]. Variations in the size and nature of ingested microplastic have also been observed in the cod stomach from different areas signifying that biota does interact with different plastics. The pervasive existence of microplastics elevates apprehensions concerning interactions between the organisms and probable contamination of the food supply to humans. This alarm has led to numerous studies on exposure and toxicity in laboratory conditions. These researches have established that varied groups of aquatic organisms be it protists, echinoderms, annelids, amphipods, cnidaria, isopods, decapods, mollusks as well as fish across different trophic levels, can consume microplastics. However, still, it is not possible to establish the facts about the trophic transfer of microplastic, because most microplastics will not translocate into the hosts' tissues. However, in aquatic organisms, both metabolic as well as negative physiological responses caused by ingestion of microplastic have only been detected in laboratory conditions after extremely high-level exposures [34]. Ingestion of microplastics has been reported from many aquatic organisms, together with economically important invertebrates and fish [31]. With the rapid increase in hotspots of micro- and meso- plastic pollution (for example: biodegradation, the release of chemicals) in different ocean depths all over the world, it is projected that ingestion of plastic has been reported in more than 690 marine and 50 freshwater organisms [95, 96, 97, 98, 99].

Among the seabird species, the first report of plastic debris ingestion in, Leach's storm petrel, *Oceanodroma leucorhoa* dates to the year 1962 [100]. Fulmars possibly represent the most studied seabirds concerning the ingestion of plastics worldwide [101]. Fulmars have been acknowledged as indicators of plastics >1mm. Surveys of substances of the gular pouch in little auks, *Alle alle* from Greenland showed concentrations of microplastic fragments and fibres to be around 9.99 particles per individual [102]. Investigations on microplastic pollution in marine mammals have also been carried out. Mammals habiting in the North Atlantic, either caught or stranded in Irish water systems showed particles throughout their digestive tracts [103]. There has been an immense increase in the manufacture of plastics since 1950s that reached almost 322 million tonnes in the year 2015, this alarming number excludes synthetic fibres which account for almost 61 million tonnes in 2015. These values are further expected to double by the year 2025 which is quite disturbing. In case of humans, the potential risk caused by ingestion of microplastic could be lowered by removing the gastrointestinal tract of most seafood species consumed [34]. However, most bivalved species and other small species of fish are consumed as a whole, which may cause exposure to microplastics [34].

Large-scale production and consumption of plastics have made a plastic web in natural habitats, which causes great damage to both biota and the economy such as consumption and entanglement of marine waste biota, introduction of non-native species, largely microorganisms, invertebrates, and seaweeds, *etc* [104, 105, 106]. Almost 373 species of wildlife were affected due to ingestion and entanglement of marine waste of which around 15 percent of species are listed in the IUCN (International Union for Conservation of Nature) red list category (Secretariat of the Convention on Biological Diversity and the Scientific and Technical Advisory Panel-GEF, 2012). Commercial effects of marine include reduced income, increased cost of cleanup, and decreased fisheries stock [107].

The production of plastic also causes the generation of greenhouse gases that have led to climate change. Microplastic particles which are frozen in Arctic Sea ice at present are expected to be released due to global warming [108]. Proper strategies for the management of waste are indispensable so as to alleviate the impacts and effects of plastic as well as microplastic pollution in aquatic habitats. Besides serving as a cause of marine remains, commercial seafood production leads to the aggregation and dispersal of microplastics in various different ways of which one of the common routes for the dispersal of plastic is discards. Pauly and Zeller (2016), reported nearly 10 million tonnes of discards each year worldwide from the year 2000 and 2010 most of which were most likely ingested by benthic scavenging species [108]. Welden and Cowie (2016) assessed the stomach contents of Clyde Sea Norway lobster and other animals belonging to the same species and other unwanted species as well, and reported the presence of microplastic contaminants [109]. Animals that contain such plastic pollutants are directly fed to other organisms, thereby leading to elevated microplastic load per individual. Frequent trawling also marks in a column of polluted sediments which makes already present plastics accessible for the filter feeders, and modifies the dispersal pattern of sediment microplastics in route with the tidal currents [110]. Offal could be another source of microplastics since a lot of fish species are usually gutted onboard during fishing processes and the remnants are usually thrown over the board. These offal are then foraged upon by the fish, seabirds and benthic organisms, so it is very much possible that the microplastics present within the offal may possibly make their entry into the food web [34].

The distribution of polystyrene microfibers and microspheres causes physiological and biological ramifications for marine organisms, particularly for filter-feeder organisms (*e.g., C. virginica, C. gigas*) that have biological importance in coastal environments, as has been already demonstrated in wild populations, laboratory sets and farmed animals as well [27, 111, 112, 113, 114]. While in the case of fish, it was also established that about eight out of the fourteen species from larval fish specimens had plastic debris in their gut contents

[115]. According to the study conducted by [106], sea turtles have a high probability of ingesting plastic debris (nearly 100%), whales have nearly 59%, seabirds (40%), while seals (~36%). On the other hand, in case of the leatherback turtle, *Dermochelys coriacea*, the first report pertaining to plastic debris ingestion dates back to 1968 [116]. In controlled experimental conditions, an assessment of sea turtles that died due to ingestion of plastic and other unrelated reasons proved that nearly half of the mortality was reported for about fourteen pieces of the plastic present in their gut [117, 118]. The models studying global distribution patterns can assess the possible rates of encounter as well as the co-occurrence of plastic pollution and organisms spatiotemporally along with the possible signs of environmental risk involved [119, 120].

It has generally been presumed that these pollutants are today present in the majority of marine, freshwater as well as terrestrial food webs, and they are perhaps the driving influences for the transfer of energy among different levels between individuals, which eventually influences the health of humans through trophic transfer [121, 122, 123].

In case of birds, *Fulmarus glacialis, Puffinus carneipes, Phoebastria immutabilis* and *Puffinus tenuirostris* groups of younger age were reported to have more microplastic contents as compared to the adult group. Asimilar trend was also observed in the franciscana dolphin, *Plontoporia blainvillei, Phoca vitulina* and Norway lobster, *Nephrops norvegicus* [106, 124, 125]. Although cases of mortality due to ingestion of plastics have been reported in numerous species such as penguins, marine turtles, seabirds, and sperm whale, however, the sublethal effects of the plastics have been greatly recognised as more noteworthy and should be thoroughly investigated across different taxonomic groups. The current scientific opinions are deduced from laboratory experimental trials that study effects on biology were because of unlikely chemical exposure, which may often lead to false conclusions, therefore, it is needed that a critical review of the metrics be used to compute ecologically appropriate and laboratory experiment trials be implemented for toxicological assays [126]. Services and applications using technologies based on space science such as remote sensing through satellite based space observations can prove to be helpful in studying both plastics as well as microplastic pollutions in the oceans, and also guide improved water management policies for the benefit of the environment and the humankind [127].

EFFECT OF MICROPLASTICS ON ORGANISMS AT CELLULAR AND SUB CELLULAR LEVEL

Microplastic as a study matter has exponentially gained attention over the last few years (GESAMP, 2015). Ingestion of microplastics has been reported to escalate free radicals (reactive oxygen species) which further leads to damage to cells and DNA [128]. Besides chemicals, the physical dimensions and properties also have an important part in ingestion as well as effects linked with particles. Another issue is the leaching of amassed and additive contaminants that could have a toxic influence. The study of the relationship between microplastics and antibiotic resistance [129, 130, 131] deciphered a new aspect that was unnoticed, and is of particular concern with respect to water cycle and could influence aquatic fauna as well. The occurrence of microplastics drastically increases the rate of recurrence of gene transfer [129]. It was suggested that the positive environment offered by microplastics is reported to augment the potential for the transfer of genes in aquatic surroundings. This could increase evolutionary progressions that would have an influence on organisms at the species level, population level and possibly community levels as well. It becomes further alarming because of the intractable nature of these microplastics that facilitates further transference potential of the antibiotic-resistant bacteria and genes over vast distances. A study [130] reported that even a simple occurrence of microplastics influences the resistance pattern of microbial communities from wastewater discharges. These workers further reported that microplastics elevated the endurance of bacterial species (WWTP-derived) and class I integron genes related to antibiotic resistance of genetic materials. They asserted that the distribution of these bacteria and related genes could prove to be a great hazard to the aquatic ecosystem. A research work [131] studied the role played by microplastics in the interaction of bacteriophages and bacteria with a specific focus on antibiotic resistance genes and reported that microplastics served as a barrier that improved the interactions between phage and bacteria other than the distribution of antibiotic resistance genes.

EFFECT OF MICROPLASTICS ON FLORA

Microplastics have been found to contaminate a wide range of aquatic environments around the world [132, 133] negatively affecting a wide range of organisms and have received much scientific attention over the last decade. There are now studies that indicated microplastics also affect flora as well [134, 20]. Therefore, in this review, an attempt has been made to highlight the possible impact of microplastics on flora. A study found that microplastics manufactured of high-density polyethylene (HDPE) and polylactic acid (PLA), and synthetic fibers can affect the development of *L. perenne*, the health of *A. rosea*, but crucial soil properties, with potential further impacts on soil ecosystem functioning [20]. De Souza Machado *et al.*, confirmed that microplastics greatly impacted the soil properties and plant performance [135]. They found significant variations in plant biomass, tissue elemental composition, root traits and soil microbial activities. Besides, their results imply that the pervasive microplastic contamination in soil may have consequences for plant performance and thus for agroecosystems and terrestrial biodiversity see Fig. (**1**). Rilling (2019), clearly demonstrated how microplastics impact plants, where they focused on the different types of microplastic, and hypothesized the effects on plants and plant community-level effects [136]. They concluded with several mechanisms through which these materials could affect plant performance. Some of them showed positive effects on roots and plant growth, others had a negative impact. These effects might vary as a function of plant species, and thus are likely to translate to changes in plant community composition and perhaps primary production. Wang *et al*, elaborated on the abundance, sources and properties of MPs in the soil, besides they analyze the combined effects of MPs and various other environmental pollutants on the soil system and also discussed the possible risks posed by MPs to soil biodiversity, food safety and human health [137]. Their results revealed that MPs can lead to combined contamination with other toxic chemicals and also confirms that packaging, plastic mulches, and sewage sludge are three major sources of soil MPs, by which they become a threat to soil fertility, food security and human health. Lozano and Rilling (2020), studied the effects of microplastic fibres and drought on plant communities [138]. Their results showed that at the community level, shoot and root mass decreased with drought, but increased with microfibers, an effect likely linked to reduced soil bulk density, improved aeration, and better penetration of roots in the soil. Moreover, their result also revealed that microfibers affected plant community structure as well. Xu *et al.*, found that microplastics not only affect the physical and chemical properties of the soil, but also damage microbial and enzyme activities, plant growth and also pose adverse ecotoxicological effects to soil fauna [139]. All these effects mainly depend on the concentration, size, and shape of microplastics, as well as soil texture. Zhou *et al.*, reviewed an article where they discuss the emerging threats of microplastics to plant and soil health. They revealed that microplastics can alter a range of key soil biogeochemical processes by changing its properties, and forming specific microbial hotspots, resulting in multiple effects on microbial activities and functions. Importantly, increased bioavailable soil carbon from the decomposition of biodegradable microplastics, which enhances microbial and enzymatic activities, potentially accelerates soil organic matter mineralization and increases nutrient competition between plants and microbes. Thus, it was suggested that biodegradable MPs appear to pose a greater risk to plant growth compared to petroleum-based MPs. Although MPs may confer some benefits in agro-ecosystems (*e.g.* enhanced soil structure, aeration). Microplastics may alter soil aggregation due to their binding to soil mineral and organic components [140]. Recently, it was revealed that microplastics can alter plant growth, soil inherent properties, and the composition and activity of microbial communities. They also investigated how the impacts of microplastic on arbuscular mycorrhizal fungi

(AMF) could alter how plants deal with other global change factors (GCFs) in the context of sustainable food production. Moreover, they presumed that co-occurrence of several GCFs, *e.g.*, elevated temperature, drought, pesticides, and microplastic could modify the impact of microplastic on AMF [141]. In one of the more recent review articles, important influences of microplastics on aquatic as well as terrestrial plants have been discussed. Their study mainly focused on the type, size and oxygen-containing group of MPs on the physical injures toward plants, which were significantly correlated to the toxicity to plants. Besides, they also reported that hydrophobic organic pollutants released from MPs showed significant chemical effects on the plants. Their study would be useful in understanding the potential toxicity of MP themselves together with released and adsorbed chemicals to plants in the environment [142].

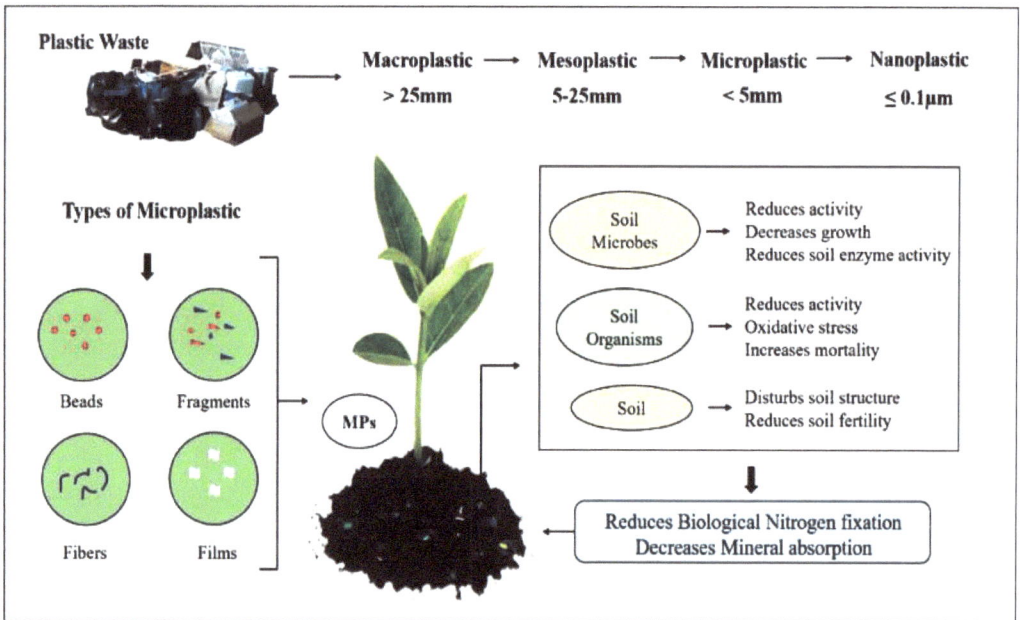

Fig. (1). Microplastic accumulation in terrestrial soil and its impact on floral community.

CONCLUSION

In this review the effects of microplastics on flora and fauna are concisely discussed. Although, a lot of work has been carried out on the effect of MPs on fauna, but comparatively less work has been done in flora so more work in this field is required. Microplastic contamination could increase in the coming future because of environmental fragmentation and disintegration of already existent wastes and further production of plastic items in the future. The probable toxic effects of microplastics on the biota have been accepted by scientific tests as validated by numerous studies carried in the past years. It is an undeniable statement that nowadays, microplastics occur and accumulate in the environment. Also, it is irrefutable that a huge number of plants as well as animals from different trophic levels are exposed to these particles and their exposure at varying concentrations could causes a number of side effects and also threatens organisms belonging to diverse species as well as the environmets they inhabit. This review will provide an insight into the existing gap in the current research works and thus could offer possible implications on the effect of microplastics on flora and fauna.

ACKNOWLEDGEMENTS

The authors are highly thankful to the Head Department of Zoology, University of Kashmir for providing necessary facilities. We are also grateful to Nafiaah Naqash, Research Scholar, Department of Life Sciences, Lovely Professional University for designing the figure.

REFERENCES

[1]   Bhatia D, Sharma NR, Singh J, Kanwar RS. Biological methods for textile dye removal from wastewater: A review. Crit Rev Environ Sci Technol 2017; 47(19): 1836-76.
[http://dx.doi.org/10.1080/10643389.2017.1393263]

[2]   Singh S, Sorokhaibam P, Kaur R, Singh R. Assessment of physicochemical parameters to investigate pollution status of white Bein: A tributary of Sutlej River. Indian J Ecol 2018; 45(4): 898-900.

[3]   Dhanjal DS, Singh SI, Bhatia DE, Singh JO, Sharma NR, Kanwar RS. Pre-treatment of the municipal wastewater with chemical coagulants. Pollut Res 2018; 37: S32-8.

[4]   Hernandez LM, Yousefi N, Tufenkji N. Are there nanoparticles in your personal care products? Environ Sci Technol Lett 2017; 4(7): 280-5.
[http://dx.doi.org/10.1021/acs.estlett.7b00187]

[5]   National Ocean Service Microplastics Available at:https://oceanservice.noaa.gov/facts/microplastics.html(2017).

[6]   Wong JKH, Lee KK, Tang KHD, Yap PS. Microplastics in the freshwater and terrestrial environments: Prevalence, fates, impacts and sustainable solutions. Sci Total Environ 2020; 719: 137512.
[http://dx.doi.org/10.1016/j.scitotenv.2020.137512] [PMID: 32229011]

[7]   Manzoor S, Kaur H, Singh R. Existence of microplastic as pollutant in harike wetland: an analysis of plastic composition and first report on ramsar wetland of india.

[8]   Zhang K, Xiong X, Hu H, *et al.* Occurrence and characteristics of microplastic pollution in Xiangxi Bay of Three Gorges Reservoir, China. Environ Sci Technol 2017; 51(7): 3794-801.
[http://dx.doi.org/10.1021/acs.est.7b00369] [PMID: 28298079]

[9]   Louisa Casson How does plastics end up in the ocean Available at:https://www.greenpeace.org.uk/plastic-end-ocean/(2017).

[10]   Naqash N, Prakash S, Kapoor D, Singh R. Interaction of freshwater microplastics with biota and heavy metals: a review. Environ Chem Lett 2020; 18(6): 1813-24.
[http://dx.doi.org/10.1007/s10311-020-01044-3]

[11]   Sapea Science Advice for Policy by European Academies consortium A Scientific Perspective on Microplastics in Nature and Society 2019.

[12]   Kaur H, Singh R. Analysis of nylon 6 as microplastic in harike wetland by comparing its ir spectra

with virgin nylon 6 and 6.6. Eur J Mol Clin Med 2020; 7(07): 2020.

[13]   Manzoor S, Naqash N, Rashid G, Singh R. Plastic material degradation and formation of microplastic in the environment: A review. Mater Today Proc 2021; 56(6): 3254-60.
[http://dx.doi.org/10.1016/j.matpr.2021.09.379]

[14]   Zhu F, Zhu C, Wang C, Gu C. Occurrence and ecological impacts of microplastics in soil systems: A review. Bull Environ Contam Toxicol 2019; 102(6): 741-9.
[http://dx.doi.org/10.1007/s00128-019-02623-z] [PMID: 31069405]

[15]   Omidi A, Naeemipoor H, Hosseini M. Plastic debris in the digestive tract of sheep and goats: an increasing environmental contamination in Birjand, Iran. Bull Environ Contam Toxicol 2012; 88(5): 691-4.
[http://dx.doi.org/10.1007/s00128-012-0587-x] [PMID: 22415646]

[16]   Zhao S, Zhu L, Li D. Microscopic anthropogenic litter in terrestrial birds from Shanghai, China: Not only plastics but also natural fibers. Sci Total Environ 2016; 550: 1110-5.
[http://dx.doi.org/10.1016/j.scitotenv.2016.01.112] [PMID: 26874248]

[17]   Souza Machado AA, Kloas W, Zarfl C, Hempel S, Rillig MC. Microplastics as an emerging threat to terrestrial ecosystems. Glob Change Biol 2018; 24(4): 1405-16.
[http://dx.doi.org/10.1111/gcb.14020] [PMID: 29245177]

[18]   Yousuf A, Naseer M, Naqash N, Singh R. Isolation and identification of microplastic particles from agricultural soil and its detection by fluorescence microscope technique. Think India Journal 2019; 22(16): 3934-49.

[19]   Boots B, Russell CW, Green DS. Effects of microplastics in soil ecosystems: Above and below ground. Environ Sci Technol 2019; 53(19): 11496-506.
[http://dx.doi.org/10.1021/acs.est.9b03304] [PMID: 31509704]

[20]   Nizzetto L, Futter M, Langaas S. Are Agricultural soils dumps for microplastics of urban origin? Environ Sci Technol 2016; 50(20): 10777-9.
[http://dx.doi.org/10.1021/acs.est.6b04140] [PMID: 27682621]

[21]   Huerta Lwanga E, Gertsen H, Gooren H, *et al.* Incorporation of microplastics from litter into burrows of Lumbricus terrestris. Environ Pollut 2017; 220(Pt A): 523-31.
[http://dx.doi.org/10.1016/j.envpol.2016.09.096] [PMID: 27726978]

[22]   Rillig MC, Ingraffia R, de Souza Machado AA. Microplastic incorporation into soil in agroecosystems. Front Plant Sci 2017; 8: 1805.
[http://dx.doi.org/10.3389/fpls.2017.01805] [PMID: 29093730]

[23]   Wegner A, Besseling E, Foekema EM, Kamermans P, Koelmans AA. Effects of nanopolystyrene on the feeding behavior of the blue mussel (*Mytilus edulis* L.). Environ Toxicol Chem 2012; 31(11): 2490-7.
[http://dx.doi.org/10.1002/etc.1984] [PMID: 22893562]

[24]   Wright SL, Rowe D, Thompson RC, Galloway TS. Microplastic ingestion decreases energy reserves in marine worms. Curr Biol 2013; 23(23): R1031-3.
[http://dx.doi.org/10.1016/j.cub.2013.10.068] [PMID: 24309274]

[25]   Köhler A. Cellular fate of organic compounds in marine invertebrates. Comp Biochem Physiol A Mol Integr Physiol 2010; 157 (Suppl.): S8.
[http://dx.doi.org/10.1016/j.cbpa.2010.06.020]

[26]   Von Moos N, Burkhardt-Holm P, Köhler A. Uptake and effects of microplastics on cells and tissue of the blue mussel *Mytilus edulis* L. after an experimental exposure. Environ Sci Technol 2012; 46(20): 11327-35.
[http://dx.doi.org/10.1021/es302332w] [PMID: 22963286]

[27]   Rochman CM, Hoh E, Kurobe T, Teh SJ. Ingested plastic transfers hazardous chemicals to fish and induces hepatic stress. Sci Rep 2013; 3(1): 3263.

[http://dx.doi.org/10.1038/srep03263] [PMID: 24263561]

[28]   Lee KW, Shim WJ, Kwon OY, Kang JH. Size-dependent effects of micro polystyrene particles in the marine copepod Tigriopus japonicus. Environ Sci Technol 2013; 47(19): 11278-83.
[http://dx.doi.org/10.1021/es401932b] [PMID: 23988225]

[29]   Sussarellu R, Soudant P, Lambert C, *et al.* Microplastics: effects on oyster physiology and reproduction. Platform presentation, International workshop on fate and impact of microplastics in marine ecosystems (MICRO2014), 13–15 January France. Plouzane 2014.

[30]   Lusher A. Microplastics in the marine environment: Distribution, interactions and effects.Marine Anthropogenic Litter. Cham: Springer 2015; pp. 245-307.
[http://dx.doi.org/10.1007/978-3-319-16510-3_10]

[31]   Duis K, Coors A. Microplastics in the aquatic and terrestrial environment: sources (with a specific focus on personal care products), fate and effects. Environ Sci Eur 2016; 28(1): 2.
[http://dx.doi.org/10.1186/s12302-015-0069-y] [PMID: 27752437]

[32]   Tang KHD. Ecotoxicological impacts of Micro and nanoplastics on Marine Fauna. Examines Mar Biol Oceanogr 2020; 3(2): 1-5.

[33]   The State of World Fisheries and Aquaculture. Contributing to food security and nutrition for all. Rome. 2016.

[34]   De Sá LC, Oliveira M, Ribeiro F, Rocha TL, Futter MN. Studies of the effects of microplastics on aquatic organisms: What do we know and where should we focus our efforts in the future? Sci Total Environ 2018; 645: 1029-39.
[http://dx.doi.org/10.1016/j.scitotenv.2018.07.207] [PMID: 30248828]

[35]   Wang W, Ndungu AW, Li Z, Wang J. Microplastics pollution in inland freshwaters of China: A case study in urban surface waters of Wuhan, China. Sci Total Environ 2017; 575: 1369-74.
[http://dx.doi.org/10.1016/j.scitotenv.2016.09.213] [PMID: 27693147]

[36]   Wright SL, Kelly FJ. Plastic and human health: a micro issue? Environ Sci Technol 2017; 51(12): 6634-47.
[http://dx.doi.org/10.1021/acs.est.7b00423] [PMID: 28531345]

[37]   Abbasi S, Soltani N, Keshavarzi B, Moore F, Turner A, Hassanaghaei M. Microplastics in different tissues of fish and prawn from the Musa Estuary, Persian Gulf. Chemosphere 2018; 205: 80-7.
[http://dx.doi.org/10.1016/j.chemosphere.2018.04.076] [PMID: 29684694]

[38]   Su L, Cai H, Kolandhasamy P, Wu C, Rochman CM, Shi H. Using the Asian clam as an indicator of microplastic pollution in freshwater ecosystems. Environ Pollut 2018; 234: 347-55.
[http://dx.doi.org/10.1016/j.envpol.2017.11.075] [PMID: 29195176]

[39]   Jabeen K, Su L, Li J, *et al.* Microplastics and mesoplastics in fish from coastal and fresh waters of China. Environ Pollut 2017; 221: 141-9.
[http://dx.doi.org/10.1016/j.envpol.2016.11.055] [PMID: 27939629]

[40]   De Sá LC, Luís LG, Guilhermino L. Effects of microplastics on juveniles of the common goby (*Pomatoschistus microps*): Confusion with prey, reduction of the predatory performance and efficiency, and possible influence of developmental conditions. Environ Pollut 2015; 196: 359-62.
[http://dx.doi.org/10.1016/j.envpol.2014.10.026] [PMID: 25463733]

[41]   Jabeen K, Li B, Chen Q, *et al.* Effects of virgin microplastics on goldfish (Carassius auratus). Chemosphere 2018; 213: 323-32.
[http://dx.doi.org/10.1016/j.chemosphere.2018.09.031] [PMID: 30237044]

[42]   Karami A, Romano N, Galloway T, Hamzah H. Virgin microplastics cause toxicity and modulate the impacts of phenanthrene on biomarker responses in African catfish (Clarias gariepinus). Environ Res 2016; 151: 58-70.
[http://dx.doi.org/10.1016/j.envres.2016.07.024] [PMID: 27451000]

[43]   Oliveira M, Ribeiro A, Hylland K, Guilhermino L. Single and combined effects of microplastics and pyrene on juveniles (0+ group) of the common goby *Pomatoschistus microps* (Teleostei, Gobiidae). Ecol Indic 2013; 34: 641-7.
[http://dx.doi.org/10.1016/j.ecolind.2013.06.019]

[44]   Avio CG, Gorbi S, Milan M, *et al.* Pollutants bioavailability and toxicological risk from microplastics to marine mussels. Environ Pollut 2015; 198: 211-22.
[http://dx.doi.org/10.1016/j.envpol.2014.12.021] [PMID: 25637744]

[45]   Qiao R, Deng Y, Zhang S, *et al.* Accumulation of different shapes of microplastics initiates intestinal injury and gut microbiota dysbiosis in the gut of zebrafish. Chemosphere 2019; 236: 124334.
[http://dx.doi.org/10.1016/j.chemosphere.2019.07.065] [PMID: 31310986]

[46]   Chen Q, Yin D, Jia Y, *et al.* Enhanced uptake of BPA in the presence of nanoplastics can lead to neurotoxic effects in adult zebrafish. Sci Total Environ 2017; 609: 1312-21.
[http://dx.doi.org/10.1016/j.scitotenv.2017.07.144] [PMID: 28793400]

[47]   Boerger CM, Lattin GL, Moore SL, Moore CJ. Plastic ingestion by planktivorous fishes in the North Pacific Central Gyre. Mar Pollut Bull 2010; 60(12): 2275-8.
[http://dx.doi.org/10.1016/j.marpolbul.2010.08.007] [PMID: 21067782]

[48]   Foekema EM, De Gruijter C, Mergia MT, van Franeker JA, Murk AJ, Koelmans AA. Plastic in north sea fish. Environ Sci Technol 2013; 47(15): 8818-24.
[http://dx.doi.org/10.1021/es400931b] [PMID: 23777286]

[49]   Yin L, Chen B, Xia B, Shi X, Qu K. Polystyrene microplastics alter the behavior, energy reserve and nutritional composition of marine jacopever (Sebastes schlegelii). J Hazard Mater 2018; 360: 97-105.
[http://dx.doi.org/10.1016/j.jhazmat.2018.07.110] [PMID: 30098534]

[50]   Sleight VA, Bakir A, Thompson RC, Henry TB. Assessment of microplastic-sorbed contaminant bioavailability through analysis of biomarker gene expression in larval zebrafish. Mar Pollut Bull 2017; 116(1-2): 291-7.
[http://dx.doi.org/10.1016/j.marpolbul.2016.12.055] [PMID: 28089550]

[51]   Nematdoost Haghi B, Banaee M. Effects of micro-plastic particles on paraquat toxicity to common carp (Cyprinus carpio): biochemical changes. Int J Environ Sci Technol 2017; 14(3): 521-30.
[http://dx.doi.org/10.1007/s13762-016-1171-4]

[52]   Barboza LGA, Vieira LR, Guilhermino L. Single and combined effects of microplastics and mercury on juveniles of the European seabass (*Dicentrarchus labrax*): Changes in behavioural responses and reduction of swimming velocity and resistance time. Environ Pollut 2018; 236: 1014-9.
[http://dx.doi.org/10.1016/j.envpol.2017.12.082] [PMID: 29449115]

[53]   Zhang S, Ding J, Razanajatovo RM, Jiang H, Zou H, Zhu W. Interactive effects of polystyrene microplastics and roxithromycin on bioaccumulation and biochemical status in the freshwater fish red tilapia (Oreochromis niloticus). Sci Total Environ 2019; 648: 1431-9.
[http://dx.doi.org/10.1016/j.scitotenv.2018.08.266] [PMID: 30340288]

[54]   Jeong CB, Kang HM, Lee MC, *et al.* Adverse effects of microplastics and oxidative stress-induced MAPK/Nrf2 pathway-mediated defense mechanisms in the marine copepod Paracyclopina nana. Sci Rep 2017; 7(1): 41323.
[http://dx.doi.org/10.1038/srep41323] [PMID: 28117374]

[55]   Cole M, Lindeque P, Fileman E, Halsband C, Galloway TS. The impact of polystyrene microplastics on feeding, function and fecundity in the marine copepod *Calanus helgolandicus*. Environ Sci Technol 2015; 49(2): 1130-7.
[http://dx.doi.org/10.1021/es504525u] [PMID: 25563688]

[56]   Liu Z, Yu P, Cai M, *et al.* Effects of microplastics on the innate immunity and intestinal microflora of juvenile *Eriocheir sinensis*. Sci Total Environ 2019; 685: 836-46.
[http://dx.doi.org/10.1016/j.scitotenv.2019.06.265] [PMID: 31247433]

[57]   Welden NAC, Cowie PR. Environment and gut morphology influence microplastic retention in langoustine, *Nephrops norvegicus.* Environ Pollut 2016; 214: 859-65.
[http://dx.doi.org/10.1016/j.envpol.2016.03.067] [PMID: 27161832]

[58]   Lusher AL, McHugh M, Thompson RC. Occurrence of microplastics in the gastrointestinal tract of pelagic and demersal fish from the English Channel. Mar Pollut Bull 2013; 67(1-2): 94-9.
[http://dx.doi.org/10.1016/j.marpolbul.2012.11.028] [PMID: 23273934]

[59]   Watts AJR, Lewis C, Goodhead RM, *et al.* Uptake and retention of microplastics by the shore crab *Carcinus maenas.* Environ Sci Technol 2014; 48(15): 8823-30.
[http://dx.doi.org/10.1021/es501090e] [PMID: 24972075]

[60]   Murray F, Cowie PR. Plastic contamination in the decapod crustacean Nephrops norvegicus (Linnaeus, 1758). Mar Pollut Bull 2011; 62(6): 1207-17.
[http://dx.doi.org/10.1016/j.marpolbul.2011.03.032] [PMID: 21497854]

[61]   Goldstein MC, Titmus AJ, Ford M. Scales of spatial heterogeneity of plastic marine debris in the northeast pacific ocean. PLoS One 2013; 8(11): e80020.
[http://dx.doi.org/10.1371/journal.pone.0080020] [PMID: 24278233]

[62]   Browne MA, Niven SJ, Galloway TS, Rowland SJ, Thompson RC. Microplastic moves pollutants and additives to worms, reducing functions linked to health and biodiversity. Curr Biol 2013; 23(23): 2388-92.
[http://dx.doi.org/10.1016/j.cub.2013.10.012] [PMID: 24309271]

[63]   Cox KD, Covernton GA, Davies HL, Dower JF, Juanes F, Dudas SE. Human consumption of microplastics. Environ Sci Technol 2019; 53(12): 7068-74.
[http://dx.doi.org/10.1021/acs.est.9b01517] [PMID: 31184127]

[64]   Revel M, Châtel A, Mouneyrac C. Micro(nano)plastics: A threat to human health? Curr Opin Environ Sci Health 2018; 1: 17-23.
[http://dx.doi.org/10.1016/j.coesh.2017.10.003]

[65]   Liu Y, Li W, Lao F, *et al.* Intracellular dynamics of cationic and anionic polystyrene nanoparticles without direct interaction with mitotic spindle and chromosomes. Biomaterials 2011; 32(32): 8291-303.
[http://dx.doi.org/10.1016/j.biomaterials.2011.07.037] [PMID: 21810539]

[66]   Awara WM, El-Nabi SH, El-Gohary M. Assessment of vinyl chloride-induced DNA damage in lymphocytes of plastic industry workers using a single-cell gel electrophoresis technique. Toxicology 1998; 128(1): 9-16.
[http://dx.doi.org/10.1016/S0300-483X(98)00008-0] [PMID: 9704901]

[67]   Hugo ER, Brandebourg TD, Woo JG, *et al.* Bisphenol A at environmentally relevant doses inhibits adiponectin release from human adipose tissue explants and adipocytes. Environ Health Perspect 2008; 116(12): 1642-7.
[http://dx.doi.org/10.1289/ehp.11537] [PMID: 19079714]

[68]   Lim SL, Ng CT, Zou L, *et al.* Targeted metabolomics reveals differential biological effects of nanoplastics and nanoZnO in human lung cells. Nanotoxicology 2019; 13(8): 1117-32.
[http://dx.doi.org/10.1080/17435390.2019.1640913] [PMID: 31272252]

[69]   Hesler M, Aengenheister L, Ellinger B, *et al.* Multi-endpoint toxicological assessment of polystyrene nano- and microparticles in different biological models *in vitro.* Toxicol. *In Vitro* 2019; 61: 104610.
[http://dx.doi.org/10.1016/j.tiv.2019.104610] [PMID: 31362040]

[70]   Wu Q, Tao H, Wong MH. Feeding and metabolism effects of three common microplastics on *Tenebrio molitor* L. Environ Geochem Health 2019; 41(1): 17-26.
[http://dx.doi.org/10.1007/s10653-018-0161-5] [PMID: 30056553]

[71]   Tourinho PS, Ivar do Sul JA, Fillmann G. Is marine debris ingestion still a problem for the coastal marine biota of southern Brazil? Mar Pollut Bull 2010; 60(3): 396-401.

[http://dx.doi.org/10.1016/j.marpolbul.2009.10.013] [PMID: 19931101]

[72] Auman HJ, Ludwig JP, Giesy JP, Colborn T. Surrey beatty & sons: albatross biology and conservation. plastic ingestion by laysan albatross chicks on sand island, midway atoll in 1994 and 1995 Available at: http://www.cosee.net/coseewest/October06Resources/ Related%20Articles/Plastic%20ingestion%20by%20Laysan%20Albatross%20chicks%20on%20Midw ay%20Atoll.pdf(1997).

[73] Hutton I, Carlile N, Priddel D. Plastic ingestion by flesh-footed shearwaters, *Puffinus carneipes*, and wedge-tailed shearwaters, *Puffinus pacificus.* Pap Proc R Soc Tasman 2008; 142(1): 67-72.
[http://dx.doi.org/10.26749/rstpp.142.1.67]

[74] Besseling E, Foekema EM, Van Franeker JA, *et al.* Microplastic in a macro filter feeder: Humpback whale *Megaptera novaeangliae.* Mar Pollut Bull 2015; 95(1): 248-52.
[http://dx.doi.org/10.1016/j.marpolbul.2015.04.007] [PMID: 25916197]

[75] Van Cauwenberghe L, Janssen CR. Microplastics in bivalves cultured for human consumption. Environ Pollut 2014; 193: 65-70.
[http://dx.doi.org/10.1016/j.envpol.2014.06.010] [PMID: 25005888]

[76] Qiu Q, Peng J, Yu X, Chen F, Wang J, Dong F. Occurrence of microplastics in the coastal marine environment: First observation on sediment of China. Mar Pollut Bull 2015; 98(1-2): 274-80.
[http://dx.doi.org/10.1016/j.marpolbul.2015.07.028] [PMID: 26190791]

[77] Zhao S, Zhu L, Wang T, Li D. Suspended microplastics in the surface water of the Yangtze Estuary System, China: First observations on occurrence, distribution. Mar Pollut Bull 2014; 86(1-2): 562-8.
[http://dx.doi.org/10.1016/j.marpolbul.2014.06.032] [PMID: 25023438]

[78] Fok L, Cheung PK. Hong Kong at the pearl river estuary: a hotspot of microplastic pollution. Mar Pollut Bull 2015; 99(1-2): 112-8.
[http://dx.doi.org/10.1016/j.marpolbul.2015.07.050] [PMID: 26233305]

[79] Dong S, Qu M, Rui Q, Wang D. Combinational effect of titanium dioxide nanoparticles and nanopolystyrene particles at environmentally relevant concentrations on nematode *Caenorhabditis elegans.* Ecotoxicol Environ Saf 2018; 161: 444-50.
[http://dx.doi.org/10.1016/j.ecoenv.2018.06.021] [PMID: 29909313]

[80] Qu M, Nida A, Kong Y, Du H, Xiao G, Wang D. Nanopolystyrene at predicted environmental concentration enhances microcystin-LR toxicity by inducing intestinal damage in *Caenorhabditis elegans.* Ecotoxicol Environ Saf 2019; 183: 109568.
[http://dx.doi.org/10.1016/j.ecoenv.2019.109568] [PMID: 31437729]

[81] Biginagwa F, Mayoma B, Shashoua Y, Syberg K, Khan F. First evidence of microplastics in the african great lakes: recovery from lake victoria nile perch and nile tilapia. J Great Lakes Res 2015.
[http://dx.doi.org/10.1016/j.jglr.2015.10.012]

[82] Horton AA, Jürgens MD, Lahive E, van Bodegom PM, Vijver MG. The influence of exposure and physiology on microplastic ingestion by the freshwater fish Rutilus rutilus (roach) in the River Thames, UK. Environ Pollut 2018; 236: 188-94.
[http://dx.doi.org/10.1016/j.envpol.2018.01.044] [PMID: 29414339]

[83] Roch S, Brinker A. Rapid and efficient method for the detection of microplastic in the gastrointestinal tract of fishes. Environ Sci Technol 2017; 51(8): 4522-30.
[http://dx.doi.org/10.1021/acs.est.7b00364] [PMID: 28358493]

[84] Hurley RR, Woodward JC, Rothwell JJ. Ingestion of microplastics by freshwater tubifex worms. Environ Sci Technol 2017; 51(21): 12844-51.
[http://dx.doi.org/10.1021/acs.est.7b03567] [PMID: 29019399]

[85] Lourenço PM, Serra-Gonçalves C, Ferreira JL, Catry T, Granadeiro JP. Plastic and other microfibers in sediments, macroinvertebrates and shorebirds from three intertidal wetlands of southern Europe and west Africa. Environ Pollut 2017; 231(Pt 1): 123-33.

[http://dx.doi.org/10.1016/j.envpol.2017.07.103] [PMID: 28797901]

[86]   Wagner M, Scherer C, Alvarez-Muñoz D, *et al*. Microplastics in freshwater ecosystems: what we know and what we need to know. Environ Sci Eur 2014; 26(1): 12.
[http://dx.doi.org/10.1186/s12302-014-0012-7] [PMID: 28936382]

[87]   Driedger AGJ, Dürr HH, Mitchell K, Van Cappellen P. Plastic debris in the laurentian great lakes: a review. J Great Lakes Res 2015; 41(1): 9-19.
[http://dx.doi.org/10.1016/j.jglr.2014.12.020]

[88]   Eerkes-Medrano D, Thompson RC, Aldridge DC. Microplastics in freshwater systems: A review of the emerging threats, identification of knowledge gaps and prioritisation of research needs. Water Res 2015; 75: 63-82.
[http://dx.doi.org/10.1016/j.watres.2015.02.012] [PMID: 25746963]

[89]   Horton AA, Walton A, Spurgeon DJ, Lahive E, Svendsen C. Microplastics in freshwater and terrestrial environments: Evaluating the current understanding to identify the knowledge gaps and future research priorities. Sci Total Environ 2017; 586: 127-41.
[http://dx.doi.org/10.1016/j.scitotenv.2017.01.190] [PMID: 28169032]

[90]   Klein S, Worch E, Knepper TP. Occurrence and spatial distribution of microplastics in river shore sediments of the Rhine-Main area in Germany. Environ Sci Technol 2015; 49(10): 6070-6.
[http://dx.doi.org/10.1021/acs.est.5b00492] [PMID: 25901760]

[91]   Free CM, Jensen OP, Mason SA, Eriksen M, Williamson NJ, Boldgiv B. High-levels of microplastic pollution in a large, remote, mountain lake. Mar Pollut Bull 2014; 85(1): 156-63.
[http://dx.doi.org/10.1016/j.marpolbul.2014.06.001] [PMID: 24973278]

[92]   Bråte ILN, Eidsvoll DP, Steindal CC, Thomas KV. Plastic ingestion by Atlantic cod (Gadus morhua) from the Norwegian coast. Mar Pollut Bull 2016; 112(1-2): 105-10.
[http://dx.doi.org/10.1016/j.marpolbul.2016.08.034] [PMID: 27539631]

[93]   Bråte ILN, Blázquez M, Brooks SJ, Thomas KV. Weathering impacts the uptake of polyethylene microparticles from toothpaste in Mediterranean mussels (M. galloprovincialis). Sci Total Environ 2018; 626: 1310-8.
[http://dx.doi.org/10.1016/j.scitotenv.2018.01.141] [PMID: 29898538]

[94]   Rochman CM, Manzano C, Hentschel BT, Simonich SLM, Hoh E. Polystyrene plastic: a source and sink for polycyclic aromatic hydrocarbons in the marine environment. Environ Sci Technol 2013; 47(24): 13976-84.
[http://dx.doi.org/10.1021/es403605f] [PMID: 24341360]

[95]   Gall SC, Thompson RC. The impact of debris on marine life. Mar Pollut Bull 2015; 92(1-2): 170-9.
[http://dx.doi.org/10.1016/j.marpolbul.2014.12.041] [PMID: 25680883]

[96]   Savoca MS, Wohlfeil ME, Ebeler SE, Nevitt GA. Marine plastic debris emits a keystone infochemical for olfactory foraging seabirds. Sci Adv 2016; 2(11): e1600395.
[http://dx.doi.org/10.1126/sciadv.1600395] [PMID: 28861463]

[97]   Galloway TS, Cole M, Lewis C. Interactions of microplastic debris throughout the marine ecosystem. Nat Ecol Evol 2017; 1(5): 0116.
[http://dx.doi.org/10.1038/s41559-017-0116] [PMID: 28812686]

[98]   Jâms IB, Windsor FM, Poudevigne-Durance T, Ormerod SJ, Durance I. Estimating the size distribution of plastics ingested by animals. Nat Commun 2020; 11(1): 1594.
[http://dx.doi.org/10.1038/s41467-020-15406-6] [PMID: 32221282]

[99]   Rothstein SI. Particle pollution of the surface of the Atlantic Ocean: Evidence from a seabird. Condor 1973; 75(3): 344-5.
[http://dx.doi.org/10.2307/1366176]

[100] Van Franeker J, Law K. Seabirds, gyres and global trends in plastic pollution. Environ Pollut 2015; 203: 89-96.

[http://dx.doi.org/10.1016/j.envpol.2015.02.034]

[101]   Amélineau F, Bonnet D, Heitz O, *et al.* Microplastic pollution in the greenland sea: background levels and selective contamination of planktivorous diving seabirds. Environ Pollut 2016; 219: 1131-9.
[http://dx.doi.org/10.1016/j.envpol.2016.09.017] [PMID: 27616650]

[102]   Lusher AL, Hernandez-Milian G, Berrow S, Rogan E, O'Connor I. Incidence of marine debris in cetaceans stranded and bycaught in Ireland: Recent findings and a review of historical knowledge. Environ Pollut 2018; 232: 467-76.
[http://dx.doi.org/10.1016/j.envpol.2017.09.070] [PMID: 28987567]

[103]   Calder D, Choong H, Carlton J, Chapman J, Miller J, Geller J. Hydroids (Cnidaria: Hydrozoa) from Japanese tsunami marine debris washing ashore in the northwestern United States. Aquat Invasions 2014; 9(4): 425-40.
[http://dx.doi.org/10.3391/ai.2014.9.4.02]

[104]   Kiessling T, Gutow L, Thiel M. Marine litter as habitat and dispersal vector.M Bergmann, L Gutow, L & M Klages. Cham, Switzerland: Springer International Publishing 2015; pp. 141-81.
[http://dx.doi.org/10.1007/978-3-319-16510-3_6]

[105]   Kühn S, Rebolledo ELB, van Franeker JA. Deleterious effects of litter on marine life. In: Bergmann M, Gutow L, Klages M, Eds. Marine Anthropogenic Litter. Springer 2015; pp. 75-116.
[http://dx.doi.org/10.1007/978-3-319-16510-3_4]

[106]   Marine plastic debris and microplastics – Global lessons and research to inspire action and guide policy change. Nairobi: United Nations Environment Programme 2016.

[107]   Obbard RW, Sadri S, Wong YQ, Khitun AA, Baker I, Thompson RC. Global warming releases microplastic legacy frozen in Arctic Sea ice. Earths Futur 2014; 2(6): 315-20.
[http://dx.doi.org/10.1002/2014EF000240]

[108]   Pauly D, Zeller D. Catch reconstructions reveal that global marine fisheries catches are higher than reported and declining. Nat Commun 2016; 7(1): 10244.
[http://dx.doi.org/10.1038/ncomms10244] [PMID: 26784963]

[109]   Welden NAC, Cowie PR. Environment and gut morphology influence microplastic retention in langoustine, *Nephrops norvegicus*. Environ Pollut 2016; 214: 859-65.
[http://dx.doi.org/10.1016/j.envpol.2016.03.067] [PMID: 27161832]

[110]   Lattin GL, Moore CJ, Zellers AF, Moore SL, Weisberg SB. A comparison of neustonic plastic and zooplankton at different depths near the southern California shore. Mar Pollut Bull 2004; 49(4): 291-4.
[http://dx.doi.org/10.1016/j.marpolbul.2004.01.020] [PMID: 15341821]

[111]   Cole M, Lindeque P, Fileman E, *et al.* Microplastic ingestion by zooplankton. Environ Sci Technol 2013; 47(12): 6646-55.
[http://dx.doi.org/10.1021/es400663f] [PMID: 23692270]

[112]   Mathalon A, Hill P. Microplastic fibers in the intertidal ecosystem surrounding Halifax Harbor, Nova Scotia. Mar Pollut Bull 2014; 81(1): 69-79.
[http://dx.doi.org/10.1016/j.marpolbul.2014.02.018] [PMID: 24650540]

[113]   Sussarellu R, Suquet M, Thomas Y, *et al.* Oyster reproduction is affected by exposure to polystyrene microplastics. Proc Natl Acad Sci USA 2016; 113(9): 2430-5.
[http://dx.doi.org/10.1073/pnas.1519019113] [PMID: 26831072]

[114]   Vázquez OA, Rahman MS. An ecotoxicological approach to microplastics on terrestrial and aquatic organisms: A systematic review in assessment, monitoring and biological impact. Environ Toxicol Pharmacol 2021; 84: 103615.
[http://dx.doi.org/10.1016/j.etap.2021.103615] [PMID: 33607259]

[115]   Carpenter EJ, Anderson SJ, Harvey GR, Miklas HP, Peck BB. Polystyrene spherules in coastal waters. Science 1972; 178(4062): 749-50.
[http://dx.doi.org/10.1126/science.178.4062.749] [PMID: 4628343]

[116] Mrosovsky N, Ryan GD, James MC. Leatherback turtles: The menace of plastic. Mar Pollut Bull 2009; 58(2): 287-9.
[http://dx.doi.org/10.1016/j.marpolbul.2008.10.018] [PMID: 19135688]

[117] Wilcox C, Puckridge M, Schuyler QA, Townsend K, Hardesty BD. A quantitative analysis linking sea turtle mortality and plastic debris ingestion. Sci Rep 2018; 8(1): 12536.
[http://dx.doi.org/10.1038/s41598-018-30038-z] [PMID: 30213956]

[118] Eriksen M, Lebreton LCM, Carson HS, *et al.* Plastic pollution in the world's oceans: More than 5 trillion plastic pieces weighing over 250,000 tons afloat at sea. PLoS One 2014; 9(12): e111913.
[http://dx.doi.org/10.1371/journal.pone.0111913] [PMID: 25494041]

[119] Schuyler QA, Wilcox C, Townsend KA, *et al.* Risk analysis reveals global hotspots for marine debris ingestion by sea turtles. Glob Change Biol 2016; 22(2): 567-76.
[http://dx.doi.org/10.1111/gcb.13078] [PMID: 26365568]

[120] Carbery M, O'Connor W, Palanisami T. Trophic transfer of microplastics and mixed contaminants in the marine food web and implications for human health. Environ Int 2018; 115: 400-9.
[http://dx.doi.org/10.1016/j.envint.2018.03.007] [PMID: 29653694]

[121] Souza Machado AA, Kloas W, Zarfl C, Hempel S, Rillig MC. Microplastics as an emerging threat to terrestrial ecosystems. Glob Change Biol 2018; 24(4): 1405-16.
[http://dx.doi.org/10.1111/gcb.14020] [PMID: 29245177]

[122] Windsor FM, Pereira MG, Tyler CR, Ormerod SJ. Persistent contaminants as potential constraints on the recovery of urban river food webs from gross pollution. Water Res 2019; 163: 114858.
[http://dx.doi.org/10.1016/j.watres.2019.114858] [PMID: 31325703]

[123] Denuncio P, Bastida R, Dassis M, Giardino G, Gerpe M, Rodríguez D. Plastic ingestion in *Franciscana dolphins, Pontoporia blainvillei* (Gervais and d'Orbigny, 1844), from Argentina. Mar Pollut Bull 2011; 62(8): 1836-41.
[http://dx.doi.org/10.1016/j.marpolbul.2011.05.003] [PMID: 21616509]

[124] Bravo Rebolledo EL, Van Franeker JA, Jansen OE, Brasseur SMJM. Plastic ingestion by harbour seals (*Phoca vitulina*) in The Netherlands. Mar Pollut Bull 2013; 67(1-2): 200-2.
[http://dx.doi.org/10.1016/j.marpolbul.2012.11.035] [PMID: 23245459]

[125] Koelmans AA, Kooi M, Law KL, van Sebille E. All is not lost: deriving a top-down mass budget of plastic at sea. Environ Res Lett 2017; 12(11): 114028.
[http://dx.doi.org/10.1088/1748-9326/aa9500]

[126] Bagchi D, Bussa R. Application of remote sensing in water quality and water resources management –an overview. 2011.

[127] Bhattacharya P, Turner JP, Ke PC. Physical adsorption of charged plastic nanoparticles affects algal photosynthesis. J Phys Chem 2010; 114: 16556-61.
[http://dx.doi.org/10.1021/jp1054759]

[128] Arias-Andres M, Klümper U, Rojas-Jimenez K, Grossart HP. Microplastic pollution increases gene exchange in aquatic ecosystems. Environ Pollut 2018; 237: 253-61.
[http://dx.doi.org/10.1016/j.envpol.2018.02.058] [PMID: 29494919]

[129] Eckert EM, Di Cesare A, Kettner MT, *et al.* Microplastics increase impact of treated wastewater on freshwater microbial community. Environ Pollut 2018; 234: 495-502.
[http://dx.doi.org/10.1016/j.envpol.2017.11.070] [PMID: 29216487]

[130] Sun M, Ye M, Jiao W, *et al.* Changes in tetracycline partitioning and bacteria/phage-comediated ARGs in microplastic-contaminated greenhouse soil facilitated by sophorolipid. J Hazard Mater 2018; 345: 131-9.
[http://dx.doi.org/10.1016/j.jhazmat.2017.11.036] [PMID: 29175125]

[131] Browne MA, Galloway T, Thompson R. Microplastic-an emerging contaminant of potential concern?

Integr Environ Assess Manag 2007; 3(4): 559-61.
[http://dx.doi.org/10.1002/ieam.5630030412] [PMID: 18046805]

[132] Lambert S, Wagner M. Microplastics are contaminants of emerging concern in freshwater environments: An overview.Freshwater Microplastics. Cham: The Handbook of Environmental Chemistry; Springer 2017; 58: pp. 1-23.

[133] Scheurer M, Bigalke M. Microplastics in Swiss floodplain soils. Environ Sci Technol 2018; 52(6): 3591-8.
[http://dx.doi.org/10.1021/acs.est.7b06003] [PMID: 29446629]

[134] Zhang GS, Liu YF. The distribution of microplastics in soil aggregate fractions in southwestern China. Sci Total Environ 2018; 642: 12-20.
[http://dx.doi.org/10.1016/j.scitotenv.2018.06.004] [PMID: 29894871]

[135] De Souza Machado AA, Lau CW, Kloas W, *et al.* Microplastics can change soil properties and affect plant performance. Environ Sci Technol 2019; 53(10): 6044-52.
[http://dx.doi.org/10.1021/acs.est.9b01339] [PMID: 31021077]

[136] Rillig MC, Lehmann A, de Souza Machado AA, Yang G. Microplastic effects on plants. New Phytol 2019; 223(3): 1066-70.
[http://dx.doi.org/10.1111/nph.15794] [PMID: 30883812]

[137] Wang J, Liu X, Li Y, *et al.* Microplastics as contaminants in the soil environment: A mini-review. Sci Total Environ 2019; 691: 848-57.
[http://dx.doi.org/10.1016/j.scitotenv.2019.07.209] [PMID: 31326808]

[138] Lozano YM, Rillig MC. Effects of Microplastic fibers and drought on plant communities. Environ Sci Technol 2020; 54(10): 6166-73.
[http://dx.doi.org/10.1021/acs.est.0c01051] [PMID: 32289223]

[139] Xu B, Liu F, Cryder Z, *et al.* Microplastics in the soil environment: Occurrence, risks, interactions and fate – A review. Crit Rev Environ Sci Technol 2020; 50(21): 2175-222.
[http://dx.doi.org/10.1080/10643389.2019.1694822]

[140] Rillig MC, Ingraffia R, de Souza Machado AA. Microplastic incorporation into soil in agroecosystems. Front Plant Sci 2017; 8: 1805.
[http://dx.doi.org/10.3389/fpls.2017.01805] [PMID: 29093730]

[141] Leifheit EF, Lehmann A, Rillig MC. Potential effects of Microplastic on arbuscular mycorrhizal fungi. Front Plant Sci 2021; 12: 626709.
[http://dx.doi.org/10.3389/fpls.2021.626709] [PMID: 33597964]

[142] Ge J, Li H, Liu P, Zhang Z, Ouyang Z, Guo X. Review of the toxic effect of microplastics on terrestrial and aquatic plants. Sci Total Environ 2021; 791: 148333.
[http://dx.doi.org/10.1016/j.scitotenv.2021.148333] [PMID: 34412379]

CHAPTER 4

Removal of Microplastic Contaminants from Aquatic Environment

Kuljit Kaur[1] and **Harpreet Kaur**[1,*]

[1] *Department of Chemistry, Lovely Professional University (LPU), Phagwara, 144411 Punjab, India*

Abstract: Microplastics (MPs) contamination has recently been recognized as a serious global concern for global food security and modern society's well-being due to its widespread presence in the aquatic and terrestrial environment. According to a growing number of reports, micro- and nanosized plastic components have been discovered in nearly every part of the world, from the bottom of the ocean to the mountain top. Microplastics have become prevalent in the environment due to the gradual disposal of plastic waste, a lack of conventional detection processes with particular removal techniques, and a slow disposal rate. By adsorbing various heavy metals, pathogens, and other chemical additives frequently utilised in the production of raw plastic, microplastics have been shown to work as potential vectors. At the tertiary level of the food chain, microplastics are consumed by marine organisms such as fish and crustaceans, and then by humans. This phenomenon is responsible for clogging digestive systems, disrupting digestion, and ultimately reducing the reproductive growth of entire living species. As a result of these repercussions, microplastics have become a growing concern as a new possible risk, demanding the management of microplastics in aquatic media. This review chapter gives a comprehensive overview of existing and newly developed technologies for detecting and removing microplastics from aquatic environments in order to minimise the ultimate possible impact on aquatic habitats.

Keywords: Activated carbon, Adsorption, Aqueous environment, Biochar, Immobilization, Microplastics, Microplastic characterization, Microplastic detection, Microplastic pollution, Microplastics remediation, Microplastics sources, Microplastic transport, Plastic waste, Primary microplastics, Porous media, Removal, Removal efficiency, Secondary microplastics, Toxicity, Water treatment.

---

* **Corresponding author Harpreet Kaur:** Department of Chemistry, Lovely Professional University (LPU), Phagwara, 144411 Punjab, India; E-mail: harpreet2.kaur@lpu.co.in

Rahul Singh and Neeta Raj Sharma (Eds.)

INTRODUCTION

Plastic pollution is a global issue that endangers both the environment and human health. Scientists and government officials have been paying close attention to the existence, distribution, and effects of plastic particles in the natural environment.

The global manufacturing of plastic has increased by 560 times in the last 60 years [1]. About half of it is used to make consumer products for one-time use, which are only used for a few moments before being discarded. Approximately 58% of plastic garbage accumulates in the natural environment, and is preserved in a solid state for the long period, whereas 18% of plastic garbage is recycled and 24% is burned [2]. The annual manufacturing of waste plastics was estimated to be greater than 348 million tons [3]. These plastics end up in the aquatic environment as a result of irresponsible handling, improper dumping, and aquaculture. When discarded plastics are subjected to sun radiation, temperature changes, waves and wind currents have physical consequences, while biological processes like fragmentation and thermal decomposition also take place [4]. Microplastics are plastic fragments having a diameter of less than 5 mm that have become the subject of research because of the danger they could cause to aquatic ecosystems and human health [5]. Microplastics are defined as plastic components, filaments, or beads with a diameter of 100 nm to <5 mm, while the particles having a diameter of less than 100 nm are called nanoplastics [6]. Based on their production pathways, they can be divided into primary and secondary microplastics [7]. Primary microplastics are commercially produced synthetically manufactured plastic pellets, beads, nurdles, fibers, and powders. Secondary microplastic particles are produced in the environment as a result of the weathering and decomposition of macroplastic and mesoplastic garbage [8, 9]. There are many ways that microplastics get into the environment, the most common of which are human activities in households, industries, and sewage systems [10]. They have the ability to adsorb heavy metals [11] and persistent organic pollutants [12] from the surrounding water environment. Organic contaminants such as polyaromatic hydrocarbons (PAHs), polychlorinated biphenyls (PCBs), perfluorinated alkyl substances (PFAS), polybrominated diethers (PBDs), and pharmaceuticals and personal care products [13], as well as trace metal contaminants such as Ag, Cd, Co, Cr, Cu, Hg, Ni, Pb, and Zn [14], can adhere to the surface of microplastics due to their hydrophobic nature. Hydrophobic organic contaminants (HOCs) have a high affinity for adsorption on non-polar surfaces including sediments and organic matter, however, they often choose surfaces made of plastic [15]. It has been outlined that if MPs and their micropollutants reach food webs *via* biota digestion, this could have consequences for ecosystems and human health [16] (Fig. **1**).

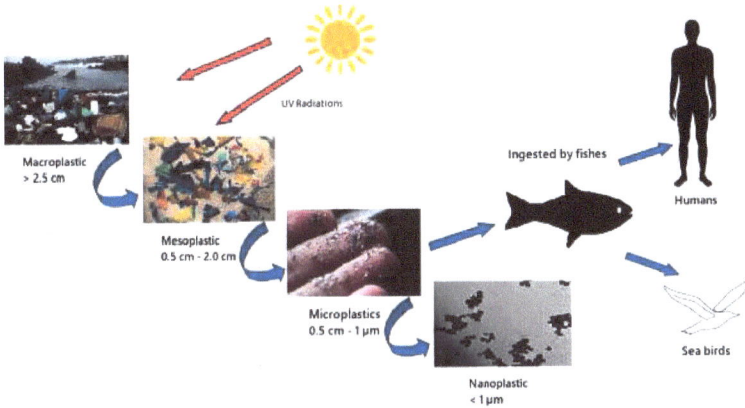

Fig. (1). Microplastics formation in the environment and their impact [17, 18, 19].

The recent global COVID-19 outbreak has increased demand for medical personal protection equipment (PPE), including single-use disposable face masks, medical face shields, and gloves, the bulk of these are composed of polymer materials [20, 21]. As a result of the pandemic, around 52 billion single-use disposable face masks were made in 2020 alone, with approximately 1.6 billion (3%) of them ending up in the oceans, where they will take 40-50 centuries to decompose. Tons of microplastics will be discharged into the oceans as a result of the natural breakdown of these masks, and these microplastics will most likely reach our food chain. Plastic waste has flooded the natural environment as a result of our excessive usage of plastic, and causing a number of environmental issues and threatening our health and survival [22, 23]. The sources, fates, occurrences, and dynamics of microplastics, as well as their interactions with biota and contaminants, have all expanded over the years. The increasing extent of microplastic waste in the environment is attracting global attention as the most serious concern addressing microplastic contamination, as well as the need for research to find long-term solutions to restrict its emissions and discharge into the environment.

In this chapter, various microplastic sources, entry paths of microplastics into the aqueous environment, and the toxicity and impacts of microplastic waste on living organisms are reviewed. We attempted to gather information on microplastic waste remediation approaches, with a focus on the removal of microplastics through adsorption and the application of biochar, a new adsorbent. In particular, biochar adsorption is examined as one of the most recent developments in the removal of microplastics.

MICROPLASTIC SOURCES AND OCCURRENCE

Microplastics (MPs) are prevalent in the environment as a solid form with a variety of dimensions, shapes, densities, colors, polymer types, and plastic particle integration with other contaminants [24]. These properties are entirely responsible for MP transportation in the ecosystem [25]. There are four different sizes of plastic waste in the environment. MPs are particles with diameters ranging from 1μm to 5 mm [26]. They can be either primary MPs generated from manufacturing and packaging, or secondary MPs, which come from mechanical impact or chemical processes fragmenting or degrading bigger particles [27]. The other three types are macroplastics (more than 20 mm), mesoplastics (5–20 mm), and nanoplastics (less than 1 m) types [28]. Fragments come in a wide range of morphologies, including fibers, microbeads, films, foams, pellets, and so on. Anyone looking for the correct mesh to capture fragments of specific sizes from water probes will find size classification to be extremely beneficial. Unfortunately, the particles chosen are not always plastic, and the challenge is to identify plastic from other types of marine waste [26].

Primary microplastics are subdivided into microbeads, microfibers, and plastic pellets depending on their shapes and applications. Very tiny spherical polymers known as microbeads are commonly found in personal care, cosmetic, and cleaning goods including toothpaste and hand sanitizers for exfoliation. They are used to encapsulate active compounds and improve product stability [29]. After usage, they are directly discharged down the drain, where they eventually end up in sewage. Polyethylene (PE) is a plastic that has been widely utilised as an exfoliating element in cosmetics and as a container material for cosmetics [30]. The residue of specific production procedures, such as abrasion or friction losses, as well as use and maintenance, may also be a source of primary microplastics. The discharge of microplastics into the environment is attributed to friction between car tyres and asphalt [31], as well as the use of synthetic clothing. Furthermore, washing textiles may cause friction and microfibers are released into the wastewater as a result of friction [32]. Microplastics have also been used in air blasting, a tool that accelerates air and abrasive material to force it through the edges and repair the surface by removing corrosion and paint. Polyester and acrylic microplastics are used as abrasive materials. This method is utilised in harbours (where tanker paint is removed) and without being treated, effluent is released into the water [33].

On the other hand, when larger raw plastic particles break down due to chemical and physical factors like pressure, climate, and UV radiation, secondary microplastics are formed [29]. Secondary microplastics are primarily found in single-use plastic bags, water bottles, and lost fishing nets [34]. These end up in the oceans, the soil, and wastewater sludge [33, 35]. Wastewater treatment plants, in particular, are a major contributor to microplastic pollution since wastewater sludge can be utilised as a soil fertiliser or thrown into the seas without being cleaned of microplastics [36]. These small plastic fragments might readily escape filtration systems, damaging the ecosystem, the food chain, and aquatic life [37]. It is worth noting that many water treatment plants are situated close to the ocean and seawater, which make them a major source of microplastic pollution. For example, out of 3340 wastewater treatment plants in mainland China, roughly 1873 (56%) are located in coastal regions, with a treatment capacity of 78106 m^3 per day, and their effluents can be discharged into aquatic environments either directly or indirectly [38]. Many studies are looking at the origin, prevalence, detection, and elimination of microplastics in water treatment plants in order to address this problem [39, 40]. The greywater filter residue had a higher polyethylene terephthalate (PET) level of 1430 mg kg^{-1} in the urban water management samples. The mass contents of the examined sewage sludge were 98.7 and 141 mg kg^{-1}, respectively, while the suspended particles of the WWTP effluent contained only 38.6 mg kg^{-1} PET [41].

Polyethylene (PE), polypropylene (PP), polyvinyl chloride (PVC), polystyrene (PS), polyethylene terephthalate (PET), polyester, acrylic, polyamides (PA), and polyurethane (PU) are the most abundant microplastics in the environment [42]. PE can be found in a variety of forms, including three different densities: linear low-density, high density variants [43, 44, 45].

Sea water, wastewater, fresh water, drinking water (both bottled and tap) food, and air, have all been shown to contain microplastics [42, 46, 47]. Plastic particles can be transported in a variety of ways, including residential wastewater discharge, sewage treatment plant emissions, stormwater runoff, and wind blowing [48, 49, 50, 51]. The primary sources of microplastics are considered to be terrestrial and freshwater environments, while the primary sinks are thought to be the sea and ocean [52]. Because macro- and microplastic particles are light, they can be carried to water bodies by the atmospheric wind. Due to their variety of shapes, tiny size, light weight, and low density, microplastics are widely transported and easily dispersed across significant distances in terrestrial and inside marine systems by storm drains, wind, along with various natural currents. Microplastics with a larger size and density sink more easily and deposit more sediment. In addition, microplastics with irregular shapes and sharp edges tend to stay submerged more frequently instead of returning to the surface, while sphere-shaped particles are more likely to return to the surface [53, 54]. Instead of being a static process, the exchange of microplastics in aquatic, terrestrial, and air ecosystems creates a dynamic recycling process of microplastic transportation in the environment [55].

TOXICITY OF MICROPLASTIC

Microplastics cause toxicity through a variety of mechanisms. There are three types of potential hazards associated with microplastics: the particles themselves that pose health risks; chemicals (unbound monomers, additives, and environmental sorbed chemicals); and biofilms, which are microorganisms that adhere to microplastics and colonise them [23].

Toxicity from Physical Properties of Microplastics

To begin with, the toxicity could be induced directly by the polymer ingredients employed in plastic product manufacturing. For example, polystyrene (PS) is commonly utilised in containers, bottles, lids, and protective packaging, however, it has been found to be able to translocate in the bloodstream and inhibit marine filter feeder reproduction [56, 57]. Also, due to their small size and pointed ends, microplastics may harm organisms and induce inflammation. Ingestion of microscopic microplastics has been shown to cause malnutrition and changes in reproduction in several organisms [58]. A study also showed that microscopic microplastics (less than 10 mm) might enter the circulatory systems of aquatic species through the gut [59].

Toxicity from Chemicals (Unbound Monomers & Additives)

Unreacted monomers, oligomers, and chemical additives released from the plastic during the long run contribute to microplastic toxicity [60]. At least one of the monomers and oligomers can both migrate from the materials used in food packaging [61]. When residual concentrations reach particular levels, humans can take them *via* a variety of routes. For example, polystyrene residues in food have been linked to major health problems. Epoxy resins containing bisphenol A, on the other hand, are absorbed by living tissues and have slow cell division rates [62].

Furthermore, chemicals like plasticizers, heat stabilizers, flame retardants, antioxidants, colourants (such as soluble azo-colorants and pigments), fillers (such as kaolin and clay), and reinforcements (such as carbon and glass fibers) added to plastics to increase their qualities may be hazardous to living organisms. Phthalates and polybrominated diphenyl ethers, for example, are two typical additives used to improve plasticity and fire resistance. They are, nevertheless, well-known as endocrine disruptive substances (EDCs). According to studies [63, 64], these substances were found in human beings and were expected to accumulate *via* bioaccumulation processes. These additives are also toxic. For instance, the quantities of bisphenol A released from food packaging materials were estimated to be between the range of 100 and 800 ng/L, whereas phthalate release levels were between g/L under the same circumstances. The majority of these additives aren't chemically bonded to the bulk plastic structures, meaning that they'll be easier to remove [65]. Hahladakis *et al.* (2018) investigated the rate of migration and release, as well as the fate of additives and their possible harmful effects on organisms and the environment. According to their study, these chemicals have the potential to migrate and cause adverse human exposure through food containers such as packaging. Various recycling and recovery techniques can also release them from plastics. As a result, safe recycling must be carried out in such a way that discharges of high-risk compounds and contamination of recycled goods are prevented at all times, assuring environmental and human health protection [66]. Oehlmann *et al.* investigated the biological impacts of frequently used plasticizers on wildlife, including dibutyl phthalate, diethylhexyl phthalate, dimethyl phthalate, butyl benzyl phthalate, and bisphenol A (BPA), with an emphasis on annelids (both aquatic and terrestrial), mollusks, crustaceans, insects, fish, and amphibians [67]. Volatile chemicals released by plastics, such as benzene, toluene, ethylbenzene, styrene, and methylene chloride, can also have long-term effects on health [68, 69]. Various additives in MPs can cause endocrine hormone synthesis to be disrupted [70].

Toxicity from Organic Pollutants, Metal-ions, & Microorganisms (Biofilm) Adsorbed on MPs

Microplastics are also efficient at adsorbing persistent organic pollutants (POPs) such as polychlorinated biphenyls (PCBs) due to their hydrophobic properties and the significant ratio of surface area to volume. They absorb hydrophobic contaminants from the water and transport them to other environments. In the North Pacific Gyre, plastic waste was collected, extracted, and studied. Some samples had high levels of polychlorinated biphenyls, PCBs, and polycyclic aromatic hydrocarbons, PAHs, and organochlorine pesticides. Because these solid-phase plastic materials float, marine creatures mistake them for food. This research shows that polluted particles are eaten and transfer toxic contaminants into the food web, posing an obvious risk to all marine species and, as a result, humans [71].

Chemical contaminants such as metals and pharmaceuticals can adsorb on microplastics and be transported up the food chain. Five different types of microplastics were used to study the adsorption of Cd, Co, Cr, Cu, Ni, Pb, and Zn by a series of batch adsorption experiments in Milli-Q water and natural fluids (seawater, urban wastewater, and irrigation water). Microplastics, particularly polyethylene and polyvinyl chloride, were found to adsorb a significant amount of lead, chromium, and zinc [72]. The ecological impact of MP and heavy metal co-exposure is mainly unknown. However, MP might move up the food chain, enhancing heavy metal toxicity and causing harm to organisms. Populations and the ecosystem as a whole may be harmed by co-exposure to heavy metals and MP Populations, communities, and the ecosystem as a whole may be harmed by co-exposure to heavy metals and MP fragments. Therefore, it is necessary to look into how MP and heavy metal co-exposure affects organisms in the food chain and future generations during the delivery process. Furthermore, MPs may travel up the food chain, enhancing the toxicity of heavy metals and causing negative effects on species. MPs may be transferred through reproduction to subsequent generations, so co-exposure to heavy metals and MP particles may have an impact on populations and communities, as well as the entire ecosystem. As a result, the possible impacts of MPs and heavy metal co-exposure on food chain organisms and the future generation in the delivery process must be explored [73].

Microplastics in the environment are readily colonised by microorganisms (live or dead), forming biofilms that influence microplastic behaviour and possible dangers [74]. Biofilms can easily cover microplastics, and the surface features of microplastics in aquatic environments influence their repercussions. Biofilms have been reported to increase the coupled toxicity of Pb (II) and microplastics [75]. Previous research has demonstrated that the presence of biofilms boosts microplastics' capacity to acquire contaminants. The combined effects of microplastics with biofilms and pollutants may be more harmful, and more research into the possible environmental dangers of coupled pollution is required [76].

MICROPLASTIC DETECTION AND CHARACTERIZATION

It is important to use sufficient identification and detection technology to efficiently remediate microplastics. There are two types of microplastic analysis: physical characterization and chemical characterization. The process of establishing the size distribution of microplastics, and other physical features including shape and color, is referred to as "physical characterization." Chemical characterization, on the other hand, was primarily utilised to analyse microplastic composition. Previously, microplastics in a particular sample were physically identified by naked-eye visual inspection. Because it is extremely dependent on the constraints of size, the sample matrix quality, and the counter, this procedure is prone to human error. As a result, this procedure is inefficient, and the findings are frequently inaccurate. As a result, new technologies and analytical procedures for the precise detection of microplastics have been developed and implemented, taking into account the complexities of microplastics and accurately identifying microplastics [77].

Stereomicroscopes are the most commonly used equipment for physical characterization. They can be used to count microplastics, measure their size, and define their shape. The size limitation imposed by the stereomicroscope's low magnification factor makes it difficult to visually identify microplastics, on the other hand, and the results are highly operator-dependent. Up to a 70% error ratio was estimated, and this error increased as particle size decreased [78]. As a result, many precautions have been taken to avoid possible mistakes. The characterization accuracy of stereomicroscopy of microplastics can be increased with the use of measurements. However, because there is no way to automate the procedure, it cannot distinguish between polymer kinds and is time-consuming [79].

Chemical characterization of microplastics can enhance the precision of microplastic identification and enable researchers to understand about their composition. Mass spectrometry and gas chromatography in combination (GC-MS), including thermal extraction desorption and pyrolysis-GC-MS, are destructive techniques [47, 80], and liquid chromatography (LC), as well as non-destructive spectroscopic techniques like FTIR spectroscopy and Raman spectroscopy, are currently used in chemical analysis. Spectroscopic techniques were the most extensively employed to evaluate microplastics in environmental samples among these methods [47, 81].

The most commonly reported approach for identifying microplastics collected in WWTPs is FTIR [37]. The variations in surface morphology and the structure of virgin and aged microplastics were examined by researchers using a field emission scanning electron microscope (SEM). A Fourier transform infrared spectrum (FTIR) obtained in the region of 4000-400 cm^{-1} was used to qualitatively detect surface functional groups of microplastics in various states. An Electroacoustic Spectrometer was utilised to analyse zeta potentials in order to measure the surface charge of microplastics [82].

The implementation of the innovative LC-UV technique, including modified sample preparation for the measurement of PET found in environmental samples, was demonstrated by Müller *et al*. The researchers detected polyethylene terephthalate (PET) signals in all environmental samples using LC-UV, including indoor dust, urban wastewater, biowaste residues, and terrestrial samples (Table **1**) [41].

Table 1. Characterization techniques for microplastic detection and analysis [71 - 75].

Characterization Technique	Characterized Property
Stereomicroscope	Number of particles, size, shape
SEM	surface morphology and structure of virgin and aged microplastics
FTIR spectroscopy	surface functional groups of microplastics in various states
Raman spectroscopy	Chemical composition of MPs
Electroacoustic Spectrometer	Surface charge on polymer
GC-MS	Chemical analysis of polymer
LC-UV	Detection of MP signals

MICROPLASTICS' REMOVAL FROM WATER

Microplastics are expected to be more harmful than other pollutants in aquatic environments because of their entanglement and bioaccumulation, resulting in the death of organisms such as fish, mammals, marine birds, and reptiles [83]. According to an ecotoxicology study [84], microplastics have been found to have harmful impacts on a variety of organisms. As a result, there is a high demand for effective microplastic removal and degradation in aqueous systems. Since microplastics are considered emergent pollutants and have been found in a variety of water sources (including freshwater, wastewater, groundwater, and oceans), researchers have been looking at several effective removal methods. Numerous methods have been used to remove MPs so far. Recent studies have investigated three key approaches to removing them from the environment: physical, chemical, and biological strategies (Fig. **2**) (Table **2**).

Out of these techniques, sorption and filtering techniques, in combination with membrane bioreactors, remove a large portion of microplastics from influents entering water treatment plants, but because effluents are released directly into aquatic environments, these systems act as daily microplastic sources. The traditional activated sludge technique is also employed in water treatment plants to treat wastewater, although it is less popular because it is less efficient than the membrane bioreactor technology. Electrocoagulation and agglomeration have been proven to be effective methods for separating microplastics, although they must be used in combination with other filtration steps. Among the aforementioned options, it has been demonstrated that photocatalytic degradation using TiO_2 and ZnO semiconductors is a feasible solution. The FTIR patterns of treated microplastics' biological degradation suggest a biological removal *via* an oxidative process. No removal technique involving organism ingestion has yet been explored, but the Red Sea giant clam has a significant potential for microplastic sorption on the shells, in addition to their breakdown in the digestive system [85].

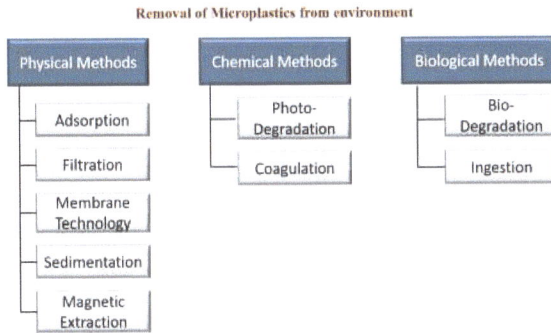

Fig. (2). Three major approaches for removal of Microplastics from environment [85, 86].

Table 2. An overview of the available research on microplastic removal techniques from the environment.

S.No.	Category	Methods	Microplastics	Material/process	Removal efficiency	References
1.	Physical Techniques	Adsorption	PE, PET, PA PS	(M-CNT)	100%	[82]
				MBC, Mg-MBC, Zn-MBC	95.02%, 94.60%, 95.79%	[83]
		Filtration	MPs	Sand Filter	29 – 44.4%	[89]
				Rapid Sand Filter	97%	[79], [90]
				Granular Activated Carbon Filtration	60%	[79], [89]
				Disc Filter	40 - 98.5%	[90]
		Sedimentation	MPs	Primary Sedimentation/ Grit Chamber	70%	[79]
				Primary Sedimentation	0.7%	[91]
		Membrane Technology	MPs	Ultrafiltration	75%	[93]
				Membrane Bioreactor	99.9%	[90]
		Magnetic Extraction	MPs	Magnetic Seeded Filtration	95%	[94]
				(M-CNT)	100%	[82]
		Dissolved Air Flotation	MPs	Conventional DAF	32.7 – 48.7%	[92]
				Posi-DAF	99%	[92]
				Tertiary Treatment	95%	[90]

(Table 2) cont.....

2.	Chemical Techniques	Coagulation	MPs PE, PET, and PA	Fe-based Salt	90.9%	[95]
			PE, PET, and PA	(FeCl3.6H2O) + Polyacrylamide (PAM)	82%, 84%, and 99%	[96]
				Alum and Aluminum Chlorohydrate	90 – 99.24%	[97]
		Electrocoagulation	MPs	Aluminum-based EC	54%	[98]
		Advanced Oxidation Processes	PE	Mn@NCNTs	72%	[99]
3.	Biological Techniques	Photo-degradation	PS and PE	TiO$_2$ Nanoparticle Film	98.40%	[100]
		Biofilter	MPs	Pilot-scale biofilter	79 – 89%	[101]
		Biodegradation	MPs	Hyper-thermophilic Bacteria	43.7%	[102]
		Ingestion	PE Microbeads	Red Sea Giant Calm	Digestive tract: 7.55 ± 1.89 beads individual^{-1} day^{-1} Shells: 66.03 ± 2.50%	[103]

Even though there is a wealth of information about biochar's ability to adsorb a wide range of compounds [104], research of the potential application of biochar in the removal of microplastics is still in its early stages. Adsorption has attracted a lot of attention in the field of pollutant removal due to its low cost, high efficiency, and simple operation technique. Several researchers recommended using porous materials to adsorb microplastics in water because of electrostatic interactions, hydrogen bond interactions, and π-π interactions between adsorbents and microplastics, and the removal efficiency was high [105]. As a result, the primary aim of this section is to study an ecologically acceptable adsorbent with a high microplastic adsorption capability. This chapter discussed a strategy for removing and degrading microplastics using biochar adsorbents, which are eco-friendly, extremely efficient, and low-cost.

MPs Removal from Water Using Magnetic Carbon Nanotubes (M-CNT)

Tang *et al.* synthesised magnetic carbon nanotubes (M-CNTs) for the first time as adsorbates for the removal of MPs. M-CNTs adhered to polyethylene (PE), polyethylene terephthalate (PET), and polyamide (PA) efficiently, and magnetic force was used to easily separate all MPs/M-CNT composites from aqueous solutions. Target MPs (5 gL^{-1}) were treated with M-CNTs (5 gL^{-1}). The agglomerates of MPs/M-CNTs were entirely eliminated within 300 minutes using magnetic forces. PE, PET, and PA had maximum adsorption capabilities of 1650, 1400, and 1100 mg-M-CNTs g^{-1}, respectively. Furthermore, they showed that the adsorbed M-CNTs can be recycled by heating at 600 °C, similar to the originals in terms of magnetic properties and capacity to remove MPs. M-CNTs continued to be effective after being used four times, eliminating 80% of all MPs in the tested solution. M-CNTs are a potential technique for MP pollution management since they are effective at removing MPs from prepared solutions and wastewater [87].

MPs Removal from Aqueous Solutions by Mg/Zn Modified Biochars (Mg/Zn-MBCs)

Wang *et al.* used Mg/Zn modified magnetic biochars to adsorb and thermally degrade microplastics from aqueous solutions. The Mg/Zn-MBCs investigated in this research exhibit an exceptional ability to adsorb onto microplastic spheres made of polystyrene suspended in water. They discovered that as a result of the added Fe, Mg, and Zn, the removal efficiencies for MBC, Mg-MBC, and Zn-MBC, respectively, are improved considerably to 94.81%, 98.75%, and 99.46%. Even after five adsorption-pyrolysis cycles, MBC (95.02%), Mg-MBC (94.60%), and Zn-MBC (95.79%) showed efficient removal of microplastics. As a result, the Mg/Zn-MBCs, which are robust, inexpensive, and eco-friendly, have a bright future in the removal of microplastics [88].

Fabrication of Chitin and Graphene Oxide (ChGO) Sponges for the Removal of Microplastics

A study on removing MPs using an adsorption technique found that a biodegradable, compressive, and porous sponge impregnated with chitin and graphene oxide (ChGO) may absorb different kinds of plastic debris. Chitin and graphene oxide (ChGO) sponges that are strong and compressible were made by Sun *et al.*, having 50 and 40 MPa of compressive stress in both the dry and wet states, respectively, for the removal of microplastics with various functional groups. Even after three adsorption-desorption cycles, because of its high adsorption capacity of 89.8%, 72.48%, and 88.9% for pure polystyrene, carboxylate-modified polystyrene, and amine-modified polystyrene, respectively, this sponge effectively adsorbed various types of MPs from water at pH 6-8 and could be reused. The electrostatic interactions, hydrogen bond interactions, and π-π interactions were the major driving forces for MPs absorption, according to the adsorption kinetic analysis, and intra-particle diffusion played a major role in the entire adsorption process. The maximal adsorption capacity of polystyrene was calculated to be 5.898, 7.528, and 8.461 mg g^{-1} at 25, 35, and 45°C, respectively, using the Langmuir isotherm model. The sponge was completely destroyed by soil microorganisms after 28 days, proving its biodegradability. Such sponges have significant potential for removing MP pollution from water due to their great reusability, biocompatibility, and biodegradability [106].

To separate PS-based MPs, Yuan *et al.* synthesised a 3D (three-dimensional) reduced graphene oxide adsorbent. The separation of microplastics from water in this experiment is mainly due to the strong π-π bond between the benzene moiety of polystyrene and the carbon atoms of 3D-reduced graphene oxide. 3D RGO has a maximum adsorption capacity of 617.28 mg/g on PS microplastics [107].

MPs Removal from Wastewater Using Pine and Spruce Bark Biochar with Steam Activation

At 475°C, by gradual pyrolysis of pine and spruce bark, Siipola *et al.* synthesized activated biochar, and steam activation was done at 800°C to alter the morphology of the biochar and improve its adsorption properties for stormwater and wastewater treatment. Before activation, spruce bark biochar had more mesopores, which were enlarged into macropores during activation. Pine bark biochar's macropore volume did not increase as a result of activation, but the mesoporosity increased ten-fold. The majority of adsorption occurs in micropores, whereas bigger pores serve as entryways into the micropores. Because bark materials exhibit mesoporosity and may be activated to create the required microporosity, they are suitable for the production of activated carbon (AC). The ability to remove different types of microplastics, such as spherical polyethylene (PE), microbeads (10 m), cylindrical PE fragments (2–3 mm), and fleece clothing fibers, was investigated. The large MP particles were completely retained; however, the absorption of MP particles at the micrometer scale was less effective. The retention of micro- and nanoplastics may be facilitated by the presence of considerably bigger pores, even if the mechanism through which microplastics sorb into biochars is not known. Physical attachment between biochar particles is most likely the mechanism for large particle retention. The better results of the spruce bark AC in the current studies, which had a limited surface area but had macro-scale porosity, is evidence for this. In order to effectively remove the smallest MP particles, higher meso-and macropore concentrations were found to be helpful [108].

Removal of Microplastic Spheres Using Corn Straw and a Hardwood Biochar

Using biochar made from corn straw and hardwood biochar as a potentially inexpensive material for incorporation in sand filter systems to enhance their capability to remove microbeads, Wang *et al.* reported research towards removing microplastic spheres by experimentation in wastewater treatment plants. Three biochar samples were tested in a leaching column at three distinct temperature ranges and the results show that biochar filters have a much higher capacity (above 95%) for removing and immobilising 10 μm diameter microplastic spheres than similar grain-sized sand filters (60-80%). According to a thorough ESEM microscopic analysis of the samples obtained following the leaching experiments, the microplastic spheres were immobilised in the samples through three morphologically controlled mechanisms known as "Stuck," "Trapped," and "Entangle" (Fig. **3**). Microplastic particles were just clogged in the sand filter, whereas microplastic spheres that remained in biochar samples were found to have all three modes of interaction. The sand filter's efficiency to remove microplastic particles is determined by the size of the sand particles and the size of the plastic particles.

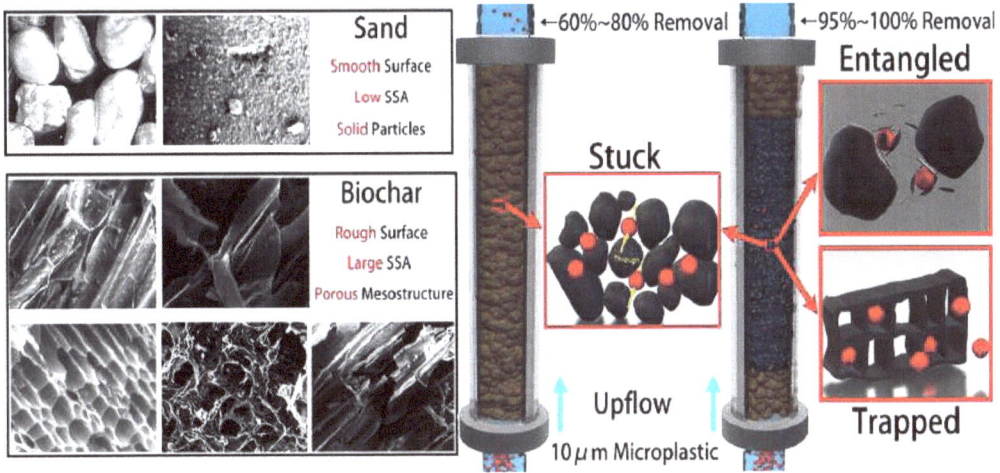

Fig. (3). Three mechanisms for microplastic immobilisation – "Stuck"; "Trapped"; "Entangled" [84].

The study explained that the first category of particle that remained in the porous system is "Stuck" in the pore system, the porous system acts like a sieve. The filter particles, which are smaller than the microplastic particles, allow the microplastic particles to accumulate in their gaps. This helps to explain why smaller microplastics predominated in water treatment plant effluent while large particles remained stuck.

Only porous filters like activated carbon and biochar have a mechanism that allows particles to be "trapped" in the pore system. When water flow is at a slow rate in pores smaller than 30 μm, plastic particles may enter the pores at a slightly bigger size and lose their mobility. Because of the 'honeycomb' and 'loofah' structures, this condition is most noticeable in hardwood biochar and C500 biochar. In comparison to sand, biochar has a much higher number of entrapped particles. Although microscopic holes can be found on the sand surface, they are rarely deep enough to prevent microplastic spheres from flowing freely. This mechanism could also account for the low removal efficiency of the C300 filter. In terms of microstructure, C300 is the biochar that comes closest to the original corn straw, and the honeycomb structure isn't the primary structure in C300 cases.

Biochar is made up of a lot of flaky particles. Particles with flaky shapes can be found in C300 and C400. Large biochar and plastic particles lack the characteristics of colloidal particles, but flaky particles do. Microplastic particles were discovered to be entangled in chips or tiny particles. The Vander Waals force would act between biochar and polystyrene particles because there is a negative charge on both surfaces. Entangled particles of microplastic would be immobilised or wrapped and attached to the filter bed due to the increase in size. In addition, by varying the SEM beam energy, it is possible to measure the thickness of flaky biochar that is entangled using interaction volume theory.

Biochar is more effective at removing contaminants than a similar-sized sand filter due to the 'Trapped' and 'Entangled' immobilization mechanisms. Biochar made from corn straw at 500°C with hardwood biochar both have a high removal and immobilization capacity due to the existence of several honeycomb patterns and thin chips. The biochars' retardation effectiveness can be summed up and ranked as: C500 > C400 > Hardwood > C300 >Silica sand based on the outcomes of the leaching column tests and with regard to the immobilisation of microplastics. This capacity could provide low-cost solutions for improving the efficiency of tertiary treatment systems and reducing microplastic discharge from WWTP outflow [109].

Nanoplastics (NPs) were efficiently removed and magnetically extracted using eco-friendly biochar modified with iron nanoparticles pyrolyzed at two temperatures: 550°C (FB-550) and 850°C (FB850), by Singh *et al*. The removal of NPs of various sizes and functionalities was demonstrated with these FB composites. The iron-modified biochar had a removal efficiency of about 100%, compared to 75% for raw biochar. The researchers proposed that nanoplastics and nanoparticles interact *via* surface complexation and electrostatic interactions controlled the adsorption processes. The adsorbent was successfully regenerated by separating nanoplastics from iron-modified biochar particles. As a result, it is a promising removal technique in the industry [110].

CONCLUSION

This brief review chapter sheds light on the occurrence, fate, detection methodologies, consequences, and removal strategies of microplastic contamination. They are all essential areas of research that require an additional focus in order to fill large knowledge gaps. To ease further investigations, testing, and treatment, it is essential to identify and detect microplastics. The growing accumulation of microplastic debris and waste needs the development of more effective plastic pollution removal and containment strategies. Biochar adsorption appears as an emerging removal approach that has been receiving a lot of scientific attention with the development of low-cost adsorption systems manufactured from biomass feedstock, such as agricultural wastes, wood chips, or municipal solid waste. Biochar has a number of advantages, including increased removal efficiency, a potentially low-cost procedure, and robust microplastic immobilization. This is especially significant when biochar is coupled with other compounds that can further entangle microplastics and increase their size, leading them to become immobile. Biochar has promising properties that make it a good choice for removing microplastics, but it comes with a number of constraints. For instance, there has been little experimental research on the topic. More research is needed to find novel and more effective techniques to generate biochar adsorbents for improved removal efficiencies as well as to look into commercialization. Furthermore, the sources and fate of microplastics must be thoroughly recognised in order to discover effective alternatives for their removal.

ACKNOWLEDGEMENTS

Authors sincerely acknowledge and express their gratitude to the Department of Chemistry, Lovely Professional University, India.

REFERENCES

[1]   Gong J, Xie P. Research progress in sources, analytical methods, eco-environmental effects, and control measures of microplastics. Chemosphere 2020; 254: 126790.
[http://dx.doi.org/10.1016/j.chemosphere.2020.126790] [PMID: 32330760]

[2]   Chamas A, Moon H, Zheng J, *et al.* Degradation rates of plastics in the environment. ACS Sustain Chem& Eng 2020; 8(9): 3494-511.
[http://dx.doi.org/10.1021/acssuschemeng.9b06635]

[3]   Shen M, Zeng G, Zhang Y, Wen X, Song B, Tang W. Can biotechnology strategies effectively manage environmental (micro)plastics?. Sci Total Environ 2019; 697: 134200.
[http://dx.doi.org/10.1016/j.scitotenv.2019.134200]

[4]   Shen M, Zhu Y, Zhang Y, *et al.* Micro(nano)plastics: Unignorable vectors for organisms. Mar Pollut Bull 2019; 139: 328-31.
[http://dx.doi.org/10.1016/j.marpolbul.2019.01.004] [PMID: 30686434]

[5]   Farrell P, Nelson K. Trophic level transfer of microplastic: Mytilus edulis (L.) to Carcinus maenas (L.). Environ Pollut 2013; 177: 1-3.
[http://dx.doi.org/10.1016/j.envpol.2013.01.046] [PMID: 23434827]

[6]   Alimi OS, Farner Budarz J, Hernandez LM, Tufenkji N. Microplastics and nanoplastics in aquatic environments: aggregation, deposition, and enhanced contaminant transport. Environ Sci Technol 2018; 52(4): 1704-24.
[http://dx.doi.org/10.1021/acs.est.7b05559] [PMID: 29265806]

[7]   Barnes DKA, Galgani F, Thompson RC, Barlaz M. Accumulation and fragmentation of plastic debris in global environments. Philos Trans R Soc Lond B Biol Sci 2009; 364(1526): 1985-98.
[http://dx.doi.org/10.1098/rstb.2008.0205] [PMID: 19528051]

[8]   Waldschläger K, Lechthaler S, Stauch G, Schüttrumpf H. The way of microplastic through the environment – Application of the source-pathway-receptor model (review). Sci Total Environ 2020; 713: 136584.
[http://dx.doi.org/10.1016/j. scitotenv.2020.136584]

[9]   Naqash N, Prakash S, Kapoor D, Singh R. Interaction of freshwater microplastics with biota and heavy metals: a review. Environ Chem Lett 2020; 18(6): 1813-24.
[http://dx.doi.org/10.1007/s10311-020-01044-3]

[10]   Sharma S, Chatterjee S. Microplastic pollution, a threat to marine ecosystem and human health: a short review. Environ Sci Pollut Res Int 2017; 24(27): 21530-47.
[http://dx.doi.org/10.1007/s11356-017-9910-8] [PMID: 28815367]

[11]   Rochman CM, Hentschel BT, Teh SJ. Long-term sorption of metals is similar among plastic types: implications for plastic debris in aquatic environments. PLoS One 2014; 9(1): e85433.
[http://dx.doi.org/10.1371/journal.pone.0085433] [PMID: 24454866]

[12]   Chua EM, Shimeta J, Nugegoda D, Morrison PD, Clarke BO. Assimilation of polybrominated diphenyl ethers from microplastics by the marine amphipod, Allorchestes compressa. Environ Sci Technol 2014; 48(14): 8127-34.
[http://dx.doi.org/10.1021/es405717z] [PMID: 24884099]

[13]   Li J, Zhang K, Zhang H. Adsorption of antibiotics on microplastics. Environ Pollut 2018; 237: 460-7.
[http://dx.doi.org/10.1016/j.envpol.2018.02.050] [PMID: 29510365]

[14]   Guzzetti E, Sureda A, Tejada S, Faggio C. Microplastic in marine organism: Environmental and toxicological effects. Environ Toxicol Pharmacol 2018; 64: 164-71.
[http://dx.doi.org/10.1016/j.etap.2018.10.009] [PMID: 30412862]

[15]   Carbery M, O'Connor W, Palanisami T. Trophic transfer of microplastics and mixed contaminants in the marine food web and implications for human health. Environ Int 2018; 115: 400-9.
[http://dx.doi.org/10.1016/j.envint.2018.03.007] [PMID: 29653694]

[16]   Browne MA, Niven SJ, Galloway TS, Rowland SJ, Thompson RC. Microplastic moves pollutants and additives to worms, reducing functions linked to health and biodiversity. Curr Biol 2013; 23(23): 2388-92.
[http://dx.doi.org/10.1016/j.cub.2013.10.012] [PMID: 24309271]

[17]   Atugoda T, Vithanage M, Wijesekara H, et al. Interactions between microplastics, pharmaceuticals and personal care products: Implications for vector transport. Environ Int 2021; 149: 106367.
[http://dx.doi.org/10.1016/j.envint.2020.106367] [PMID: 33497857]

[18]   Bush E. Microplastics in the human body: what we know and don't know. NBC News, NBC news 2022. Available at:https://www.nbcnews.com/science/science-news/microplastics-human-body--now-dont-know-rcna23331(2022).

[19]   Lehner R, Weder C, Petri-Fink A, Rothen-Rutishauser B. Emergence of nanoplastics in the environment and possible impact on human health. Environ Sci Technol 2019; 53(4): 1748-65.
[http://dx.doi.org/10.1021/acs.est.8b05512] [PMID: 30629421]

[20]   Mehran MT, Raza Naqvi S, Ali Haider M, Saeed M, Shahbaz M, Al-Ansari T. Global plastic waste management strategies (Technical and behavioral) during and after COVID-19 pandemic for cleaner global urban life. Energy Sources A Recovery Util Environ Effects 2021; 1-10.
[http://dx.doi.org/10.1080/15567036.2020.1869869]

[21]   Ahmad I, Ahmed I, Naqash N, Mehmood S. Novel coronavirus (Covid-19) a ubiquitous hazard to human health: a review. Journal of Ecophysiology and Occupational Health 2020; 20(3&4): 185-95.
[http://dx.doi.org/10.18311/jeoh/2020/25381]

[22] Zhang K, Hamidian AH, Tubić A, *et al.* Understanding plastic degradation and microplastic formation in the environment: *A review.* Environ Pollut 2021; 274: 116554.
[http://dx.doi.org/10.1016/j.envpol.2021.116554] [PMID: 33529891]

[23] Singh R, Manzoor S, Naqash N. Microplastic hazard, management, remediation, and control strategies: a review. Int J Environ Technol Manag 2022; 1(1): 10049175.
[http://dx.doi.org/10.1504/IJETM.2022.10049175]

[24] Manzoor S, Kaur H, Singh R. Existence of Microplastic as Pollutant in Harike Wetland: An Analysis of Plastic Composition and First Report on Ramsar Wetland of India.

[25] Lambert S, Scherer C, Wagner M. Ecotoxicity testing of microplastics: Considering the heterogeneity of physicochemical properties. Integr Environ Assess Manag 2017; 13(3): 470-5.
[http://dx.doi.org/10.1002/ieam.1901] [PMID: 28440923]

[26] Hartmann NB, Hüffer T, Thompson RC, *et al.* Are we speaking the same language? Recommendations for a definition and categorization framework for plastic debris. Environ Sci Technol 2019; 53(3): 1039-47.
[http://dx.doi.org/10.1021/acs.est.8b05297] [PMID: 30608663]

[27] Sharma S, Chatterjee S. Microplastic pollution, a threat to marine ecosystem and human health: a short review. Environ Sci Pollut Res Int 2017; 24(27): 21530-47.
[http://dx.doi.org/10.1007/s11356-017-9910-8] [PMID: 28815367]

[28] Napper IE, Thompson RC. Plastic debris in the marine environment: history and future challenges. Glob Chall 2020; 4(6): 1900081.
[http://dx.doi.org/10.1002/gch2.201900081] [PMID: 32685195]

[29] Coyle R, Hardiman G, Driscoll KO. Microplastics in the marine environment: A review of their sources, distribution processes, uptake and exchange in ecosystems. Case Studies in Chemical and Environmental Engineering 2020; 2: 100010.
[http://dx.doi.org/10.1016/j.cscee.2020.100010]

[30] Lee S, Lee TG. A novel method for extraction, quantification, and identification of microplastics in CreamType of cosmetic products. Sci Rep 2021; 11(1): 18074.
[http://dx.doi.org/10.1038/s41598-021-97557-0] [PMID: 34508145]

[31] Lassen C, Hansen SF, Magnusson K, *et al.* 2015). Microplastics: Occurrence, effects and sources of releases to the environment in Denmark. Danish Environmental Protection Agency.
http://mst.dk/service/publikationer/publikationsarkiv/2015/nov/rapport-ommikroplast

[32] De Falco F, Di Pace E, Cocca M, Avella M. The contribution of washing processes of synthetic clothes to microplastic pollution. Sci Rep 2019; 9(1): 6633.
[http://dx.doi.org/10.1038/s41598-019-43023-x] [PMID: 31036862]

[33] Boucher J, Friot D. Sources Primary Microplastics in the Oceans: a Global Evaluation of Sources. Gland, Switzerland: IUCN 2017.
[http://dx.doi.org/10.2305/IUCN.CH.2017.01.en]

[34] Montarsolo A, Mossotti R, Patrucco A, *et al.* Study on the microplastics release from fishing nets. Eur Phys J Plus 2018; 133(11): 494.
[http://dx.doi.org/10.1140/epjp/i2018-12415-1]

[35] Yousuf A, Naseer M, Naqash N, Singh R. Isolation and Identification of microplastic particles from agricultural soil and its detection by fluorescence microscope technique. Think India Journal 2019; 22(16): 3934-49.

[36] Sharma S, Chatterjee S. Microplastic pollution, a threat to marine ecosystem and human health: a short review. Environ Sci Pollut Res Int 2017; 24(27): 21530-47.
[http://dx.doi.org/10.1007/s11356-017-9910-8] [PMID: 28815367]

[37] Murphy F, Ewins C, Carbonnier F, Quinn B. Wastewater treatment works (WwTW) as a source of

microplastics in the aquatic environment. Environ Sci Technol 2016; 50(11): 5800-8.
[http://dx.doi.org/10.1021/acs.est.5b05416] [PMID: 27191224]

[38]   Jin L, Zhang G, Tian H. Current state of sewage treatment in China. Water Res 2014; 66(66): 85-98.
[http://dx.doi.org/10.1016/j.watres.2014.08.014] [PMID: 25189479]

[39]   Sun J, Dai X, Wang Q, van Loosdrecht MCM, Ni BJ. Microplastics in wastewater treatment plants:
Detection, occurrence and removal. Water Res 2019; 152: 21-37.
[http://dx.doi.org/10.1016/j.watres.2018.12.050] [PMID: 30660095]

[40]   Carr SA, Liu J, Tesoro AG. Transport and fate of microplastic particles in wastewater treatment plants.
Water Res 2016; 91: 174-82.
[http://dx.doi.org/10.1016/j.watres.2016.01.002] [PMID: 26795302]

[41]   Müller A, Goedecke C, Eisentraut P, Piechotta C, Braun U. Microplastic analysis using chemical
extraction followed by LC-UV analysis: a straightforward approach to determine PET content in
environmental samples. Environ Sci Eur 2020; 32(1): 85.
[http://dx.doi.org/10.1186/s12302-020-00358-x]

[42]   Manzoor S, Kaur H, Singh R. Analysis Of nylon 6 as microplastic in harike wetland by comparing its
ir spectra with virgin nylon 6 and 6.6. Eur J Mol Clin Med 2020; 7(07): 2020.

[43]   Padervand M, Lichtfouse E, Robert D, Wang C. Removal of microplastics from the environment. A
review. Environ Chem Lett 2020; 18(3): 807-28.
[http://dx.doi.org/10.1007/s10311-020-00983-1]

[44]   Gewert B, Plassmann MM, MacLeod M. Pathways for degradation of plastic polymers floating in the
marine environment. Environ Sci Process Impacts 2015; 17(9): 1513-21.
[http://dx.doi.org/10.1039/C5EM00207A] [PMID: 26216708]

[45]   Fotopoulou KN, Karapanagioti HK. Degradation of various plastics in the environment. Hazardous
chemicals associated with plastics in the marine environment. Cham: Springer International Publishing
2019; pp. 71-92.

[46]   Koelmans AA, Mohamed Nor NH, Hermsen E, Kooi M, Mintenig SM, De France J. Microplastics in
freshwaters and drinking water: Critical review and assessment of data quality. Water Res 2019; 155:
410-22.
[http://dx.doi.org/10.1016/j.watres.2019.02.054] [PMID: 30861380]

[47]   Dris R, Imhof HK, Löder MGJ, Gasperi J, Laforsch C, Tassin B. Microplastic contamination in
freshwater systems: methodological challenges, occurrence and sources. Microplastic Contamination
in Aquatic Environments 2018; pp. 51-93.
[http://dx.doi.org/10.1016/B978-0-12-813747-5.00003-5]

[48]   Mason SA, Garneau D, Sutton R, *et al.* Microplastic pollution is widely detected in US municipal
wastewater treatment plant effluent. Environ Pollut 2016; 218: 1045-54.
[http://dx.doi.org/10.1016/j.envpol.2016.08.056] [PMID: 27574803]

[49]   Nizzetto L, Bussi G, Futter MN, Butterfield D, Whitehead PG. A theoretical assessment of
microplastic transport in river catchments and their retention by soils and river sediments. Environ Sci
Process Impacts 2016; 18(8): 1050-9.
[http://dx.doi.org/10.1039/C6EM00206D] [PMID: 27255969]

[50]   Dris R, Gasperi J, Mirande C, *et al.* A first overview of textile fibers, including microplastics, in
indoor and outdoor environments. Environ Pollut 2017; 221: 453-8.
[http://dx.doi.org/10.1016/j.envpol.2016.12.013] [PMID: 27989388]

[51]   Siegfried M, Koelmans AA, Besseling E, Kroeze C. Export of microplastics from land to sea. A
modelling approach. Water Res 2017; 127: 249-57.
[http://dx.doi.org/10.1016/j.watres.2017.10.011] [PMID: 29059612]

[52]   Manzoor S, Naqash N, Rashid G, Singh R. Plastic material degradation and formation of microplastic
in the environment: A review. Mater Today Proc 2021.

[53]   Horton AA, Svendsen C, Williams RJ, Spurgeon DJ, Lahive E. Large microplastic particles in sediments of tributaries of the River Thames, UK – Abundance, sources and methods for effective quantification. Mar Pollut Bull 2017; 114(1): 218-26.
[http://dx.doi.org/10.1016/j.marpolbul.2016.09.004] [PMID: 27692488]

[54]   Kowalski N, Reichardt AM, Waniek JJ. Sinking rates of microplastics and potential implications of their alteration by physical, biological, and chemical factors. Mar Pollut Bull 2016; 109(1): 310-9.
[http://dx.doi.org/10.1016/j.marpolbul.2016.05.064] [PMID: 27297594]

[55]   Xiang Y, Jiang L, Zhou Y, *et al.* Microplastics and environmental pollutants: Key interaction and toxicology in aquatic and soil environments. J Hazard Mater 2022; 422: 126843.
[http://dx.doi.org/10.1016/j.jhazmat.2021.126843] [PMID: 34419846]

[56]   Sussarellu R, Suquet M, Thomas Y, *et al.* Oyster reproduction is affected by exposure to polystyrene microplastics. Proc Natl Acad Sci USA 2016; 113(9): 2430-5.
[http://dx.doi.org/10.1073/pnas.1519019113] [PMID: 26831072]

[57]   Chen J, Tan M, Nemmar A, *et al.* Quantification of extrapulmonary translocation of intratracheal-instilled particles *in vivo* in rats: Effect of lipopolysaccharide. 2006; 222(3): 195-201.
[http://dx.doi.org/10.1016/j.tox.2006.02.016]

[58]   Besseling E, Wang B, Lürling M, Koelmans AA. Nanoplastic affects growth of S. obliquus and reproduction of *D. magna*. Environ Sci Technol 2014; 48(20): 12336-43.
[http://dx.doi.org/10.1021/es503001d] [PMID: 25268330]

[59]   Browne MA, Dissanayake A, Galloway TS, Lowe DM, Thompson RC. Ingested microscopic plastic translocates to the circulatory system of the mussel, Mytilus edulis (L). Environ Sci Technol 2008; 42(13): 5026-31.
[http://dx.doi.org/10.1021/es800249a] [PMID: 18678044]

[60]   Thompson RC, Olsen Y, Mitchell RP, *et al.* Lost at sea: where is all the plastic? Science 2004; 304(5672): 838.
[http://dx.doi.org/10.1126/science.1094559] [PMID: 15131299]

[61]   Piringer OG, Baner AL. Plastic packaging materials for food: barrier function, mass transport, quality assurance, and legislation. Hoboken: Wiley 2008.
[http://dx.doi.org/10.1002/9783527621422]

[62]   Lau OW, Wong SK. Contamination in food from packaging material. J Chromatogr A 2000; 882(1-2): 255-70.
[http://dx.doi.org/10.1016/S0021-9673(00)00356-3] [PMID: 10895950]

[63]   Talsness CE, Andrade AJM, Kuriyama SN, Taylor JA, vom Saal FS. Components of plastic: experimental studies in animals and relevance for human health. Philos Trans R Soc Lond B Biol Sci 2009; 364(1526): 2079-96.
[http://dx.doi.org/10.1098/rstb.2008.0281] [PMID: 19528057]

[64]   Teuten E L, Saquing J M, Knappe D R U, *et al.* Transport and release of chemicals from plastics to the environment and to wildlife 2009; 364(1526): 2027-45.
[http://dx.doi.org/10.1098/rstb.2008.0284]

[65]   Fasano E, Bono-Blay F, Cirillo T, Montuori P, Lacorte S. Migration of phthalates, alkylphenols, bisphenol A and di(2-ethylhexyl)adipate from food packaging. Food Control 2012; 27(1): 132-8.
[http://dx.doi.org/10.1016/j.foodcont.2012.03.005]

[66]   Hahladakis JN, Velis CA, Weber R, Iacovidou E, Purnell P. An overview of chemical additives present in plastics: Migration, release, fate and environmental impact during their use, disposal and recycling. J Hazard Mater 2018; 344: 179-99.
[http://dx.doi.org/10.1016/j.jhazmat.2017.10.014] [PMID: 29035713]

[67]   Oehlmann J, Schulte-Oehlmann U, Kloas W, *et al.* A critical analysis of the biological impacts of plasticizers on wildlife. Philos Trans R Soc Lond B Biol Sci 2009; 364(1526): 2047-62.

[http://dx.doi.org/10.1098/rstb.2008.0242] [PMID: 19528055]

[68] Huff J, Chan P, Melnick R. Clarifying carcinogenicity of ethylbenzene. Regul Toxicol Pharmacol 2010; 58(2): 167-9.
[http://dx.doi.org/10.1016/j.yrtph.2010.08.011] [PMID: 20723573]

[69] Wexler P, Gad SC. Encyclopedia of toxicology. New Yok: Academic Press 1998.

[70] Liu G, Zhu Z, Yang Y, Sun Y, Yu F, Ma J. Sorption behavior and mechanism of hydrophilic organic chemicals to virgin and aged microplastics in freshwater and seawater. Environ Pollut 2019; 246: 26-33.
[http://dx.doi.org/10.1016/j.envpol.2018.11.100] [PMID: 30529938]

[71] Rios LM, Jones PR, Moore C, Narayan UV. Quantitation of persistent organic pollutants adsorbed on plastic debris from the Northern Pacific Gyre's "eastern garbage patch". J Environ Monit 2010; 12(12): 2226-36.
[http://dx.doi.org/10.1039/c0em00239a] [PMID: 21042605]

[72] Godoy V, Blázquez G, Calero M, Quesada L, Martín-Lara MA. The potential of microplastics as carriers of metals. Environ Pollut 2019; 255(Pt 3): 113363.
[http://dx.doi.org/10.1016/j.envpol.2019.113363] [PMID: 31614247]

[73] Cao Y, Zhao M, Ma X, *et al.* A critical review on the interactions of microplastics with heavy metals: Mechanism and their combined effect on organisms and humans. Sci Total Environ 2021; 788: 147620.
[http://dx.doi.org/10.1016/j.scitotenv.2021.147620] [PMID: 34029813]

[74] Wang J, Guo X, Xue J. Biofilm-developed microplastics as vectors of pollutants in aquatic environments. Environ Sci Technol 2021; 55(19): acs.est.1c04466.
[http://dx.doi.org/10.1021/acs.est.1c04466] [PMID: 34553907]

[75] Qi K, Lu N, Zhang S, Wang W, Wang Z, Guan J. Uptake of Pb(II) onto microplastic-associated biofilms in freshwater: Adsorption and combined toxicity in comparison to natural solid substrates. J Hazard Mater 2021; 411: 125115.
[http://dx.doi.org/10.1016/j.jhazmat.2021.125115] [PMID: 33486230]

[76] He S, Jia M, Xiang Y, *et al.* Biofilm on microplastics in aqueous environment: Physicochemical properties and environmental implications. J Hazard Mater 2022; 424(Pt B): 127286.
[http://dx.doi.org/10.1016/j.jhazmat.2021.127286] [PMID: 34879504]

[77] Shim WJ, Hong SH, Eo S. Identification methods in microplastic analysis: A review. Anal Methods 2016.
[http://dx.doi.org/10.1039/C6AY02558G]

[78] Hidalgo-Ruz V, Gutow L, Thompson RC, Thiel M. Microplastics in the marine environment: a review of the methods used for identification and quantification. Environ Sci Technol 2012; 46(6): 3060-75.
[http://dx.doi.org/10.1021/es2031505] [PMID: 22321064]

[79] Carr SA, Liu J, Tesoro AG. Transport and fate of microplastic particles in wastewater treatment plants. Water Res 2016; 91(91): 174-82.
[http://dx.doi.org/10.1016/j.watres.2016.01.002] [PMID: 26795302]

[80] Dümichen E, Eisentraut P, Bannick CG, Barthel AK, Senz R, Braun U. Fast identification of microplastics in complex environmental samples by a thermal degradation method. Chemosphere 2017; 174: 572-84.
[http://dx.doi.org/10.1016/j.chemosphere.2017.02.010] [PMID: 28193590]

[81] Elert AM, Becker R, Duemichen E, *et al.* Comparison of different methods for MP detection: What can we learn from them, and why asking the right question before measurements matters? Environ Pollut 2017; 231(Pt 2): 1256-64.
[http://dx.doi.org/10.1016/j.envpol.2017.08.074] [PMID: 28941715]

[82] Ye S, Cheng M, Zeng G, *et al.* Insights into catalytic removal and separation of attached metals from

natural-aged microplastics by magnetic biochar activating oxidation process. Water Res 2020; 179: 115876.
[http://dx.doi.org/10.1016/j.watres.2020.115876] [PMID: 32387922]

[83]   Cole M, Lindeque P, Halsband C, Galloway TS. Microplastics as contaminants in the marine environment: A review. Mar Pollut Bull 2011; 62(12): 2588-97.
[http://dx.doi.org/10.1016/j.marpolbul.2011.09.025] [PMID: 22001295]

[84]   Wu M, Yang C, Du C, Liu H. Microplastics in waters and soils: Occurrence, analytical methods and ecotoxicological effects. Ecotoxicol Environ Saf 2020; 202: 110910.
[http://dx.doi.org/10.1016/j.ecoenv.2020.110910] [PMID: 32800245]

[85]   Padervand M, Lichtfouse E, Robert D, Wang C. Removal of microplastics from the environment. A review. Environ Chem Lett 2020; 18(3): 807-28.
[http://dx.doi.org/10.1007/s10311-020-00983-1]

[86]   Bhatt P, Pathak VM, Bagheri AR, Bilal M. Microplastic contaminants in the aqueous environment, fate, toxicity consequences, and remediation strategies. Environ Res 2021; 200: 111762.
[http://dx.doi.org/10.1016/j.envres.2021.111762] [PMID: 34310963]

[87]   Tang Y, Zhang S, Su Y, Wu D, Zhao Y, Xie B. Removal of microplastics from aqueous solutions by magnetic carbon nanotubes. Chem Eng J 2021; 406: 126804.
[http://dx.doi.org/10.1016/j.cej.2020.126804]

[88]   Wang J, Sun C, Huang QX, Chi Y, Yan JH. Adsorption and thermal degradation of microplastics from aqueous solutions by Mg/Zn modified magnetic biochars. J Hazard Mater 2021; 419: 126486.
[http://dx.doi.org/10.1016/j.jhazmat.2021.126486] [PMID: 34214855]

[89]   Wang Z, Lin T, Chen W. Occurrence and removal of microplastics in an advanced drinking water treatment plant (ADWTP). Sci Total Environ 2020; 700: 134520.
[http://dx.doi.org/10.1016/j.scitotenv.2019.134520] [PMID: 31669914]

[90]   Talvitie J, Mikola A, Koistinen A, Setälä O. Solutions to microplastic pollution – Removal of microplastics from wastewater effluent with advanced wastewater treatment technologies. Water Res 2017; 123: 401-7.
[http://dx.doi.org/10.1016/j.watres.2017.07.005] [PMID: 28686942]

[91]   Liu X, Yuan W, Di M, Li Z, Wang J. Transfer and fate of microplastics during the conventional activated sludge process in one wastewater treatment plant of China. Chem Eng J 2019; 362: 176-82.
[http://dx.doi.org/10.1016/j.cej.2019.01.033]

[92]   Zhang Z, Su Y, Zhu J, Shi J, Huang H, Xie B. Distribution and removal characteristics of microplastics in different processes of the leachate treatment system. Waste Manag 2021; 120: 240-7.
[http://dx.doi.org/10.1016/j.wasman.2020.11.025] [PMID: 33310600]

[93]   Rhein F, Scholl F, Nirschl H. Magnetic seeded filtration for the separation of fine polymer particles from dilute suspensions: Microplastics. Chem Eng Sci 2019; 207: 1278-87.
[http://dx.doi.org/10.1016/j.ces.2019.07.052]

[94]   Wang Y, Li Y, Tian L, Ju L, Liu Y. The removal efficiency and mechanism of microplastic enhancement by positive modification dissolved air flotation. Water Environ Res 2021; 93(5): 693-702.
[http://dx.doi.org/10.1002/wer.1352] [PMID: 32363675]

[95]   Ma B, Xue W, Ding Y, Hu C, Liu H, Qu J. Removal characteristics of microplastics by Fe-based coagulants during drinking water treatment. J Environ Sci (China) 2019; 78: 267-75.
[http://dx.doi.org/10.1016/j.jes.2018.10.006] [PMID: 30665645]

[96]   Lapointe M, Farner JM, Hernandez LM, Tufenkji N. Understanding and improving microplastic removal during water treatment: impact of coagulation and flocculation. Environ Sci Technol 2020; 54(14): 8719-27.
[http://dx.doi.org/10.1021/acs.est.0c00712] [PMID: 32543204]

[97]   Perren W, Wojtasik A, Cai Q. Removal of microbeads from wastewater using electrocoagulation. ACS Omega 2018; 3(3): 3357-64.
[http://dx.doi.org/10.1021/acsomega.7b02037] [PMID: 31458591]

[98]   Kang J, Zhou L, Duan X, Sun H, Ao Z, Wang S. Degradation of cosmetic microplastics *via* functionalized carbon nanosprings. Matter 2019; 1(3): 745-58.
[http://dx.doi.org/10.1016/j.matt.2019.06.004]

[99]   Ariza-Tarazona MC, Villarreal-Chiu JF, Hernández-López JM, *et al.* Microplastic pollution reduction by a carbon and nitrogen-doped TiO$_2$: Effect of pH and temperature in the photocatalytic degradation process. J Hazard Mater 2020; 395: 122632.
[http://dx.doi.org/10.1016/j.jhazmat.2020.122632] [PMID: 32315794]

[100] Nabi I, Bacha AUR, Li K, *et al.* Complete photocatalytic mineralization of microplastic on TiO$_2$ nanoparticle film. iScience 2020; 23(7): 101326.
[http://dx.doi.org/10.1016/j.isci.2020.101326] [PMID: 32659724]

[101] Liu F, Nord N, Bester K, Vollertsen J. Microplastics removal from treated wastewater by a biofilter. Water 2020; 12(4): 1085.
[http://dx.doi.org/10.3390/w12041085]

[102] Chen Z, Zhao W, Xing R, *et al.* Enhanced *in situ* biodegradation of microplastics in sewage sludge using hyperthermophilic composting technology. J Hazard Mater 2020; 384: 121271.
[http://dx.doi.org/10.1016/j.jhazmat.2019.121271] [PMID: 31611021]

[103] Arossa S, Martin C, Rossbach S, Duarte CM. Microplastic removal by Red Sea giant clam (Tridacna maxima). Environ Pollut 2019; 252(Pt B): 1257-66.
[http://dx.doi.org/10.1016/j.envpol.2019.05.149] [PMID: 31252123]

[104] Dai Y, Zhang N, Xing C, Cui Q, Sun Q. The adsorption, regeneration and engineering applications of biochar for removal organic pollutants: A review. Chemosphere 2019; 223: 12-27.
[http://dx.doi.org/10.1016/j.chemosphere.2019.01.161] [PMID: 30763912]

[105] Sun C, Wang Z, Chen L, Li F. Fabrication of robust and compressive chitin and graphene oxide sponges for removal of microplastics with different functional groups. Chem Eng J 2020; 393: 124796.
[http://dx.doi.org/10.1016/j.cej.2020.124796]

[106] Sun C, Wang Z, Chen L, Li F. Fabrication of robust and compressive chitin and graphene oxide sponges for removal of microplastics with different functional groups. Chem Eng J 2020; 393: 124796.
[http://dx.doi.org/10.1016/j.cej.2020.124796]

[107] Yuan F, Yue L, Zhao H, Wu H. Study on the adsorption of polystyrene microplastics by three-dimensional reduced graphene oxide. Water Sci Technol 2020; 81(10): 2163-75.
[http://dx.doi.org/10.2166/wst.2020.269] [PMID: 32701494]

[108] Siipola V, Pflugmacher S, Romar H, Wendling L, Koukkari P. Low-cost biochar adsorbents for water purification including microplastics removal. Appl Sci (Basel) 2020; 10(3): 788.
[http://dx.doi.org/10.3390/app10030788]

[109] Wang Z, Sedighi M, Lea-Langton A. Filtration of microplastic spheres by biochar: removal efficiency and immobilisation mechanisms. Water Res 2020; 184: 116165.
[http://dx.doi.org/10.1016/j.watres.2020.116165] [PMID: 32688153]

[110] Singh N, Khandelwal N, Ganie ZA, Tiwari E, Darbha GK. Eco-friendly magnetic biochar: An effective trap for nanoplastics of varying surface functionality and size in the aqueous environment. Chem Eng J 2021; 418: 129405.
[http://dx.doi.org/10.1016/j.cej.2021.129405]

CHAPTER 5

Status of Microplastic Pollution in Natural Water Bodies

Sadguru Prakash[1,*]

¹ Department of Zoology, M.L.K. (P.G.) College, Balrampur, UP, India

Abstract: The presence of microplastics in the environment has been declared as an emerging pollutant because the production of plastic is increasing tremendously throughout the world without proper management. Microplastics (MPs) are small plastic particles (size <5mm) released directly from the use of cosmetic products, or indirectly through the degradation of large plastic items under environmental conditions. Nowadays, it is estimated that annually between 4 and 14 million tonnes of plastic go into the seas and are hazardous to aquatic life. Fishes may ingest microplastics either directly or from the prey containing these particles. MPs were found between the stomach, gut, and intestine of the fish. These MPs accumulated in the fish body which causes serious health issues leading to mortality of the fish. MPs can cause various eco-toxicological effects on fish like behavioral change, cytotoxicity, neurotoxicity effects, liver stress, *etc.*

Keywords: Abundance, Bioaccumulation, Environment, Eco-toxicity, Freshwater, Fish, Hazard, Health-risk, Human, Interaction, Microplastic, Occurrence, Pollutant, Primary, Secondary, Toxicity.

INTRODUCTION

The word plastic originally referred to any substance that was easily molded and shaped (from the Greek adjective 'plastikos'). Plastics were originally developed well before the twentieth century using natural materials such as insect secretion shellac, latex from tree sap, rubber, and celluloid. However, today when referring to plastics, we tend to mean synthetic long-chain organic polymers derived from the polymerization of monomers extracted from petroleum other products, including polyvinyl chloride (PVC), nylon, polyethylene (PE), polystyrene (PS), and poly-propylene (PP) [1].

* **Corresponding author Sadguru Prakash:** Department of Zoology, M.L.K.P.G. College, Balrampur, U.P. India;
E-mail: sadguruprakash@gmail.com

Rahul Singh and Neeta Raj Sharma (Eds.)

Common plastic polymers include PP, PE, low-density polyethylene (LDPE), and polyacrylates [2]. They are lightweight, inexpensive, and durable materials, which can easily be sculptured into a variety of products that retrieve use in an extensive application. Plastics have attained a crucial status in modern life and are now ubiquitous [3]. Environmental conditions like thermooxidative process, photo oxidative method, ultrasonic processes, and microorganisms cause degradation of plastics into small fragments [4]. When plastic particles reach upto the size of <5mm, they are categorized as microplastics. The term "Microplastic" was formally introduced by Thompson [5].

Nowadays, microplastic particles have been ubiquitously detected in almost all aquatic habitats like deep oceans, river, lakes and sediments of the planet, in a broad range of shapes, polymers, sizes and concentrations in the environments of marine water, freshwater, agroecosystems, atmosphere, food and drinking-water, biota, and other remote locations [6]. Microplastics consist of carbon and hydrogen atoms bound together in polymer chains. Other chemicals, such as phthalates, polybrominated diphenyl ethers (PBDEs), and tetrabromobisphenol A (TBBPA), are typically also present in microplastics, and many of these chemical additives leach out of the plastics after entering the environment. Microplastics are actually a group of different toxins with varying hues and shapes [7]. Microplastics can contain two types of chemicals: (i) additives and polymeric raw materials (*e.g.*, monomers or oligomers) originating from the plastics, and (ii) chemicals absorbed from the surrounding ambience. Lubricants and anti-adhesives are substances that facilitate the processing of plastic materials, improving their flow characteristics.

Microplastics found in the environment are a very heterogeneous group of particles differing in size, shape, chemical composition and specific density that originate from a variety of different sources. Based on their origin, they are also categorized into primary and secondary microplastics depending on whether the particles were originally manufactured to be that size (primary) or whether they have resulted from the breakdown of macroplastics (secondary). Thus primary microplastics are small sized plastic particles or fragments that are less than 5 mm in size before releasing directly into the environment whereas secondary microplastics are the fragmentation of larger plastic materials' degradation under environmental conditions [8].

Examples of primary microplastics include microbeds found in personal care products, plastic pellets (or nurdles) used in industrial manufacturing, and plastic fibres used in synthetic textiles (*e.g.*, nylon). Primary microplastics enter the environment directly through any of various channels for example, product use (*e.g.*, personal care products being washed into wastewater systems from households), unintentional loss from spills during manufacturing or transport, or abrasion during washing (*e.g.*, laundering of clothing made with synthetic textiles). Secondary microplastics form from the breakdown or degradation of larger plastics products (such as water or soda bottles and plastic bags) when they enter the environment; this typically happens when larger plastics undergo weathering, through exposure to, for example, wave action, wind abrasion, and ultraviolet radiation from sunlight. Microplastics are not biodegradable so when these enter once in the environment either in the form of primary or secondary microplastics. Microplastics also are a source of air pollution, occurring in dust and airborne fibrous particles. The health effects of microplastics inhalation are unknown.

Microplastic can be also categorized by their form, commonly in fibers, fragments, and spherical beads, as well as by their chemical composition, for example, polyethylene (PE), low-density PE (LDPE), PE terephthalate (PET), polyacrylates (PA), and so on [9]. MPs can be divided into many groups depending on the characteristics considered, describing a diversified class of materials that include a wide range of polymer types, particle sizes (ranging over 6 orders of magnitude), shapes (from spheres to fibers), and chemical formulations (thousands of different types), which are likely to be found in water [10]. Microplastic are of special concern science their bioaccumulation potential increases with decreasing size. MPs may be ingested by various organisms ranging from plankton and fish to birds and even mammals, and accumulate throughout the aquatic food web. Although plastics can absorb organic contaminants from the surrounding media. Since these compounds can transfer to organisms upon ingestion and serve as vectors for other organic pollutants and are, therefore, a source of exposure to organisms for these chemicals [11].

SOURCES OF MICROPLASTICS INTO THE ENVIRONMENT

The world's production of plastics (*i.e.* synthetic organic polymers) has strongly expanded during the last decades, from 1.7 million tons in 1950 to 299 million tons in 2013 [12]. Plastics have been found virtually in all environments ranging from the arctic to deserts to household dust. They are mainly introduced into the environment through ineffective waste management practices. Under environmental influences such as ultraviolet light and physical abrasion, the larger plastic particles degrade into macrocplastics (> 25 mm) which are degraded into mesoplastics (5-25mm) and then into microplastics (>5mm) in diameter [11]. Thus, microplastics, small pieces of plastic (<5mm), occur in the environment as a consequence of plastic pollution. The gradual reduction in size facilitates the transfer of plastic to a longer distance. By this, plastic can be considered a major emerging pollutant globally. These particles can transport other harmful chemicals used as additives in their fabrication or accumulate on them due to strong adsorption capacity of microplastics [13]. Moreover, the adsorption capability of microplastic increases with a reduction in size [11]. Recently the term nanoplastics are introduced for small microplastic particles ranging in size from 0.2-2mm (commonly <100 nm in any of their dimensions). However not all MPs (secondary MPs) are the result of degradation of larger particles.

Many primary MPs are released into the environment in the form of microbeads, resin pellets or personal care products (PCPs) [11], particularly from domestic wastewater. Thus microplastics are small plastic particles released directly from the use of cosmetic products, or indirectly through the degradation of large plastic items like bags, bottle and synthetic cloths also under environmental conditions. Many of these products readily enter the environment in wastes. Runoff from urban, agricultural, and recreational activities, indiscriminate disposal (plastic bottles, packaging, and shipping), industrial release (including fisheries and cosmetics), atmospheric fallout, and WWTP effluents are among the multiple plastic and microplastic sources to the environment. Raw materials used for the fabrication of plastic products (pre-production plastics), namely plastic resin pellets or flakes and plastic powder or fluff, are another important source of primary microplastics. They can reach the environment after accidental loss during transport or with run-off from processing facilities, *i.e.* often as a result of improper handling [12].

Microplastics have been found in a variety of environments, including oceans and freshwater ecosystems [14]. In oceans alone, annual plastic pollution, from all types of plastics, was estimated at 4 million to 14 million tons in the early 21st century [15]. Microplastics are also used in medical applications, *e.g.* in dentist tooth polish, and as carriers to deliver active pharmaceutical agents [16, 17]. These are also used in industrial abrasives *i.e.* for air-blasting to remove paint from metal surfaces and for cleaning the engines and machines [16, 18]. Industrial abrasives contain *e.g.* acrylic, PS, melamine, polyester (PES) and poly allyl diglycol carbonate microplastics [19]. After use, microplastics from personal care products, medical products and in industrial abrasives can reach the environment *via* wastewater.

Primary MPs are those already manufactured with a micro size, including the microspheres (<500 µm) such as pellets and microbeads contained in some cosmetic products, mixtures used for sandblasting/shot-blasting, and MPs employed as pharmaceuticals vectors and to form 3D printing [9]. It was estimated that the main contributors of primary microplastic emission to surface water are cleaning agents, paints, coatings and cosmetic products [13]. Microplastic particles are released directly into the environment directly from the use of specific personal care or cosmetic products such as hand cleaners, facial cleaners and toothpaste [17]. Skin cleaners contain microparticles like polyolefin particles (74–420 µm in size) in the form of polyethylene (PE), polypropylene (PP) and polystyrene (PS). Gouin *et al.* (2015) [20] estimated that in 2012, approx. 6% of the liquid skin cleaning products contained microplastics. They also accounted that 93% of the microplastics used in skin cleaning products in the form of polypropylene (PP).

Secondary MPs are the products of degradation of larger plastic materials, from mechanical or photo-oxidative pathways see Table **1**. These are introduced by the disintegration or decomposition of larger plastic material fragments (macroplastics) after entering into the environment by intense weathering, exposure to ultraviolet radiation, mechanical forces, thermal degradation, photolysis, thermo-oxidation and bio-degradation processes [21]. Secondary MPs arising by washing clothes are generally polyester, acrylic, and polyamide which can be more than 100 fibers per litre of effluent [22, 23].

Table 1. Feasible degradation course of synthetic polymers [27].

Specification	Thermal / Thermo-oxidative Degradation	Photo-degradation	Biodegradation
Vital ingredients	UV-light / high energy radiation	Heat and /or Oxygen	Micro-organisms
Heat obligation	None	Above ambient temperature	None
Degradation scale	Slow initiation, fast propagation	Fast	Moderate

Moreover, synthetic textiles are also an important source of microplastics. Synthetic textile fibres are released to water from waste water of domestic washing machines and in air and dust, either during normal use or during tumble drying. In addition, synthetic fibres are released into the environment from hygiene products, *e.g.* if improperly disposed into wastewater [12].

The other sources of microplastics are abrasion from car tyros, ship paints and other protective paints that contain synthetic polymers, *e.g.* alkyds, poly (acrylate/styrene), PU and epoxy resins. Microplastics may be released as a consequence of abrasion from household plastics materials, by spills during application of the paint, by abrasion during the use of the painted product and during paint removal [16].

Microplastics persist in the environment at high levels in aquatic and marine ecosystem. Plastic degrades slowly which takes hundreds of years. The probability of accumulation of microplastics in the bodies and tissues of many organisms increased in 2014 from 15 to 51 trillion individual pieces of microplastic that are in the world's oceans. Their weight was estimated to be between 93000 and 236000 metric tons.

IMPACT OF MICROPLASTICS IN FRESHWATER FISH

There are numerous ways through which MPs and associated contaminants get incorporated into the aquatic biota. These include filter feeding, suspension feeding, inhalation at air-water surface and consumption of prey exposed to MPs or through direct ingestion. Ingestion is believed to be the main MPs exposure route for several aquatic animals. Aquatic animals including plankton passively ingest MPs due to their inability to differentiate MPs and food.

The effects of MPs contamination on fish health are not yet fully understood. Fishes may ingest microplastics either directly or by the prey containing these particles [24]. Pinheiro *et al.,* (2017) [25] concluded that, 34 species of fresh water fishes were found sensitive throughout the world. Raven *et al.,* (2020) [26] observed microplastics in all 49 fish species inhabiting in two freshwater reservoirs of Bloomington city of Illinois and reported that microplastics were more concentrated in the guts rather than gills. Although these numbers are very low, it may be due to the paucity of research on the accumulation and impact of MPs in freshwater fishes. Due to strong adsorption capacity of microplastics, they provide surface area for various bio-organic or inorganic toxic substances; the ingestion of these adsorbed toxin containing MPs could be a serious health issue for fish.

The minute size, buoyancy and attractive color of MPs particles make them ideal candidates as food for fish [27]. Within aquatic ecosystems, microplastics can have quite a harmful effect on local fish fauna which are contaminated through the ingestion of MPs. The ingestion of MPs by fish can result in their accumulation in the digestive tract of fish which can cause starvation because of the false sensation of satiation or even perforation of the gastrointestinal tract. These MPs have negative physical and physiological effects on fish [28].

The negative physical effects include clogging and inflammation of the digestive system and laceration of gastrointestinal tissues which disturb the mechanism of absorption of nutrients [29]. The physiological interference can also be observed when MPs directly interfere with the immune system of fish through the stimulation of degranulation and through behavioural change, reducing the ability of a predator to perceive [28, 30]. Internal and digestive enzyme system may get damaged and even the reproduction system because of MPs digestion [31, 32]. Tiny particles of low-density polyethylene (LDPE) were exposed to environmental bay condition for consecutive three months and then fed to fish. Soon after two months, the tissues of fish had a greater concentration of PBTs and showed signs of liver stress, glycogen depletion, fatty vacuolation and cell necrosis [33].

Jabeen *et al.* (2017) [34] studied the relationship between plastic pollution and feeding traits as well as habitats of freshwater fishes and observed that fish inhabiting freshwater waterbodies of urban areas were under the higher risk of MPs exposure. MPs were ingested more frequently in these fishes [35]. A similar observation was also made by some workers who reported that fishes collected from river near urbanized areas showed a significantly higher proportion of ingestion of plastic debris in relation to fish caught in less urbanized areas [36, 37]. They did not observe MPs in edible freshwater fish collected from upstream areas, while those collected from urban rivers had MPs in their gut, supporting the hypothesis that wastewater treatment plants, in urbanized areas, are one of the sources of MPs in inland surface waters. Raza and Khan, (2018) [38] concluded that in fishes, MPs cause the reduction in the feeding activity, oxidative stress, genotoxicity, neurotoxicity, retardation in growth, reduction in reproductive fitness and ultimately death.

Thus it can be concluded that microplastics are consumed by fishes *via* a variety of methods and cause adverse effects leading to mortality, neurotoxicity, cytotoxicity, liver stress, behavioural changes, oxidative stress, genotoxicity, *etc.* Plastic abundance was found between the stomach, gut, and intestine of the fishes.

ENVIRONMENTAL AND HEALTH IMPACTS OF MICOPLASTICS

Burning of plastics in the open air leads to environmental pollution due to the release of poisonous chemicals. The polluted air, when inhaled by humans and animals, affects their health and can cause respiratory problems. Microplastics have been detected in drinking water, beer, and food products, including seafood and table salt see Fig. (**1**). In a pilot study involving eight individuals from eight different countries, microplastics were recovered from stool samples of every participant. Scientists have also detected microplastics in human tissues and organs. The implications of these findings for human health were uncertain.

Microplastics are not easily seen by the naked eye but are found in many areas, including lakes, rivers, wetlands [8], oceans, sea ice, remote islands, Antarctic's, sediments and soil [39], as well as in the digestive systems, respiratory structures, and tissues of wildlife, including birds, mammals, reptiles, fish and shellfish [40]. Aquatic organisms ingest MPs that accumulate through the food chain more readily than larger plastics. Their bioaccumulation potential increases with decreasing size [41, 42]. In the environment, MPs and NPs may undergo various transformations commonly associated with natural or anthropogenic colloids, namely, homo- and hetero-aggregation, interactions with microorganisms and macromolecules (*e.g.*, adsorption of proteins, natural organic matter), and biodegradation [43, 44].

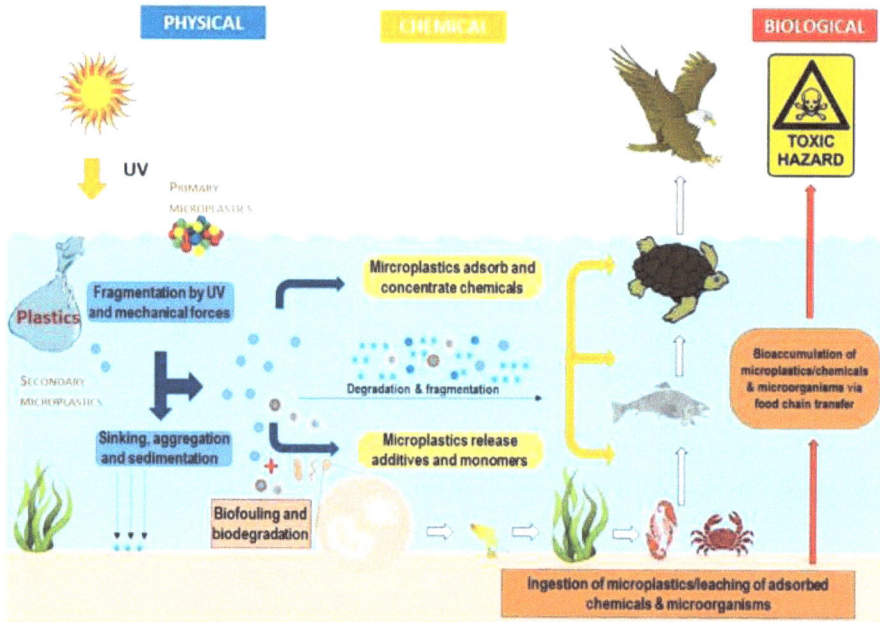

Fig. (1). Physical, chemical, and biological processes affecting MPs in the aquatic environment.

Microplastics have been found lodged in the digestive tracts and tissues of various invertebrate sea animals, including crustaceans such as crabs. Fish and birds are likely to ingest microplastics floating on the water surface, mistaking the plastic bits for food. It is demonstrated that the ingestion of MPs by aquatic biota increases the bioaccumulation of plastic additives. Ingested microplastic particles can physically damage organs and leach hazardous chemicals from the hormone-disrupting bisphenol A (BPA) to pesticides that can compromise immune function and stymie growth and reproduction. The ingestion of microplastics can cause aquatic species to consume less food and therefore, to have less energy to carry out life functions, and it can result in neurological and reproductive toxicity.

EFFECTS OF MICROPLASTIC ON HUMAN HEALTH

WHO (2019) emphasized the ubiquitous presence of microplastics in the environment and also its effects on human health. The major quantity of microplastic enter into the human body by the ingestion of contaminated food [32]. Human beings intake microplastics about 80 g per day *via* fruits and vegetables which accumulate MPs through the uptake from polluted soil [45]. 0.44 MPs/g, 0.11 MPs/g, 0.03 MPs/g, and 0.09 MPs/g were found in sugar, salt, alcohol, and bottled water, respectively [46]. The microplastic has been found between 3 to 5 fibers per 10 g of mussels of marine animal food (fish, bivalves and crustaceans) [47]. When these microplastics enter a human body *via* diet, synthetic particles smaller than 150 μm cross the gastrointestinal epithelium in mammalian bodies, which causes systemic exposure.

Human health risks are mainly attributed to the chemical additives on plastics and sorbed toxic compounds present on microplastics. Monomers of certain plastic polymers such as ethylene and ethylene terephthalate are not recognized as a significant human health threat but styrene exhibits estrogen like activity [48] and vinyl chloride is a genetoxic and mutagenic agent [49]. Chemical additives added to plastics *viz.*, phthalates, bisphenol-A, nonylphenols and flame retardants act as endocrine disruptors [50]. Toxic chemicals sorbed to microplastics and harmful pathogens attached to the microplastics have the potential to negatively affect the human health. When marine organisms ingest plastic debris, these contaminants enter their digestive systems, and overtime accumulate in the food web. The transfer of contaminants between marine species and humans through the consumption of seafood has been identified as a health hazard.

SUGGESTIONS TO REDUCE THE MICROPLASTIC POLLUTION

1. Reduce using single-use plastics: Where you live, the easiest and the most direct way that you can get started is by reducing your own use of single-use plastics. Single-use plastics include plastic bags, water bottles, straws, cups, utensils, dry cleaning bags, take-out containers, and any other plastic items that are used once and then discarded.

The best way to do this is by (a) refusing any single-use plastics that you do not need (*e.g.* straws, plastic bags, takeout utensils, takeout containers), and (b) purchasing, and carrying with you, reusable versions of those products, including reusable grocery bags, produce bags, bottles, utensils, coffee cups, and dry cleaning garment bags. And when you refuse single-use plastic items, help businesses by letting them know that you would like them to offer alternatives.

2. Buy and use the plastic-free cosmetics.

3. Recycle the plastic properly.

4. Buy clothes made from natural materials.

CONCLUSION

Microplastics found in marine and freshwater environments around the world, are raising concerns about the long term impact on animals and ecosystems in addition to recent discoveries of MPs entering the human food chain. As plastics take more than 300 years to disintegrate into micro and nano-plastics, this indicates the impact of plastic pollution on aquatic animal health and human health for years to come. The occurrence of microplastics in fish species used for human consumption is a global problem and we are vulnerable to microplastic exposure through the consumption of seafood and other human food items, as well as through other routes such as air. Based on the review of literature, the following conclusions can be drawn:

1. Regulate some rules to counter the generation of MPs in water bodies.

2. There should be a ban or monitoring over the production of personal care products containing MPs, as they are one of the primary sources.

3. Toxic effects and biomagnifications of MPs through food chains need to be evaluated comprehensively.

ACKNOWLEDGEMENT

Declared none.

REFERENCES

[1]   Vert M, Doi Y, Hellwich KH, *et al.* Terminology for biorelated polymers and applications (IUPAC Recommendations 2012). Pure Appl Chem 2012; 84(2): 377-410.
[http://dx.doi.org/10.1351/PAC-REC-10-12-04]

[2]   Frias JPGL, Otero V, Sobral P. Evidence of microplastics in samples of zooplankton from Portuguese coastal waters. Mar Environ Res 2014; 95: 89-95.
[http://dx.doi.org/10.1016/j.marenvres.2014.01.001] [PMID: 24461782]

[3]   Halden RU. Plastics and health risks. Annu Rev Public Health 2010; 31(1): 179-94.
[http://dx.doi.org/10.1146/annurev.publhealth.012809.103714] [PMID: 20070188]

[4]   Ballabio C, Panagos P, Lugato E, *et al.* Copper distribution in European topsoils: An assessment based on LUCAS soil survey. Sci Total Environ 2018; 636: 282-98.
[http://dx.doi.org/10.1016/j.scitotenv.2018.04.268] [PMID: 29709848]

[5]   Thompson RC, Olsen Y, Mitchell RP, *et al.* Lost at sea: where is all the plastic?. Science 2004; 304: 838.
[http://dx.doi.org/10.1126/science.1094559]

[6]   Singh R, Manzoor S, Naqash N. Microplastic hazard, management, remediation, and control strategies: a review. Int J Environ Technol Manag 2022; 1(1): 10049175.
[http://dx.doi.org/10.1504/IJETM.2022.10049175]

[7]   Rochman CM, Brookson C, Bikker J, *et al.* Rethinking microplastics as a diverse contaminant suite. Environ Toxicol Chem 2019; 38(4): 703-11.
[http://dx.doi.org/10.1002/etc.4371] [PMID: 30909321]

[8]   Manzoor S, Kaur H, Singh R. Existence of microplastic as pollutant in harike wetland: an analysis of plastic composition and first report on ramsar wetland of india. Curr World Environ 2021; 16(1): 123-33.
[http://dx.doi.org/10.12944/CWE.16.1.12]

[9]   Silva AB, Bastos AS, Justino CIL, da Costa JP, Duarte AC, Rocha-Santos TAP. Microplastics in the environment: Challenges in analytical chemistry - A review. Anal Chim Acta 2018; 1017: 1-19.
[http://dx.doi.org/10.1016/j.aca.2018.02.043] [PMID: 29534790]

[10]   Jiang JQ. Occurrence of microplastics and its pollution in the environment: A review. Sustainable Production and Consumption 2018; 13: 16-23.
[http://dx.doi.org/10.1016/j.spc.2017.11.003]

[11]   Wagner M, Scherer C, Alvarez-Muñoz D, *et al.* Microplastics in freshwater ecosystems: what we know and what we need to know. Environ Sci Eur 2014; 26(1): 12.
[http://dx.doi.org/10.1186/s12302-014-0012-7] [PMID: 28936382]

[12]   Duis K, Coors A. Microplastics in the aquatic and terrestrial environment: sources (with a specific focus on personal care products), fate and effects. Environ Sci Eur 2016; 28(1): 2.
[http://dx.doi.org/10.1186/s12302-015-0069-y] [PMID: 27752437]

[13]   Naqash N, Prakash S, Kapoor D, Singh R. Interaction of freshwater microplastics with biota and heavy metals: a review. Environ Chem Lett 2020; 18(6): 1813-24.
[http://dx.doi.org/10.1007/s10311-020-01044-3]

[14]   Manzoor S, Singh R, Kaur H. Microplastic in marine organism: Environmental and toxicological effects. Envir Toxi and Pharm 2021; 64: 164-71.
[http://dx.doi.org/10.1016/j.etap.2018.10.009]

[15]   Manzoor S, Naqash N, Rashid G, Singh R. Plastic material degradation and formation of microplastic in the environment: a review. Mater Today Proc 2022; 56: 3254-60.
[http://dx.doi.org/10.1016/j.matpr.2021.09.379]

[16]   Sundt P, Schulze P-E, Syversen F. Sources of microplastic-pollution to the marine environment. Mepex Nor Environ Agency 2014; 86: 20.

[17]   Lassen C, Hansen SF, Magnusson K, *et al.* Microplastics-Occurrence, effects and sources of. Significance 2012; 2.

[18]   Essel R, Engel L, Carus M, Ahrens RH. Sources of microplastics relevant to marine protection in Germany. Texte 2015; 64: 1219-26.

[19]   Eriksen M, Mason S, Wilson S, *et al.* Microplastic pollution in the surface waters of the Laurentian Great Lakes. Mar Pollut Bull 2013; 77(1-2): 177-82.
[http://dx.doi.org/10.1016/j.marpolbul.2013.10.007] [PMID: 24449922]

[20]   Gouin T, Avalos J, Brunning I, *et al.* Use of micro-plastic beads in cosmetic products in Europe and their estimated emissions to the North Sea environment. SOFW J 2015; 141: 40-6.

[21]   Zhao S, Zhu L, Li D. Microplastic in three urban estuaries, China. Environ Pollut 2015; 206: 597-604.
[http://dx.doi.org/10.1016/j.envpol.2015.08.027] [PMID: 26312741]

[22]   Habib D, Locke DC, Cannone LJ. Synthetic fibers as indicators of municipal sewage sludge, sludge products, and sewage treatment plant effluents. Water Air Soil Pollut 1998; 103(1/4): 1-8.
[http://dx.doi.org/10.1023/A:1004908110793]

[23]   Browne MA, Crump P, Niven SJ, *et al.* Accumulation of microplastic on shorelines woldwide: sources and sinks. Environ Sci Technol 2011; 45(21): 9175-9.
[http://dx.doi.org/10.1021/es201811s] [PMID: 21894925]

[24]   Desforges JPW, Galbraith M, Dangerfield N, Ross PS. Widespread distribution of microplastics in subsurface seawater in the NE Pacific Ocean. Mar Pollut Bull 2014; 79(1-2): 94-9.
[http://dx.doi.org/10.1016/j.marpolbul.2013.12.035] [PMID: 24398418]

[25]   Pinheiro C, Oliveira U, Vieira M. Occurrence and impacts of microplastics in freshwater fish. J Aquac Mar Biol 2017; 5: 138.

[26]   Hurt R, O'Reilly CM, Perry WL. Microplastic prevalence in two fish species in two U.S. reservoirs.

Limnol Oceanogr Lett 2020; 5(1): 147-53.
[http://dx.doi.org/10.1002/lol2.10140]

[27]   Kumar S, Rajesh M, Rajesh KM, Suyani NK, Rasheeq Ahamed A, Pratiksha KS. Impact of microplastics on aquatic organisms and human health: A review. Int J Environ Sci Nat Resour 2020; 26: 59-64.

[28]   Lönnstedt OM, Eklöv P. RETRACTED: Environmentally relevant concentrations of microplastic particles influence larval fish ecology. Science 2016; 352(6290): 1213-6.
[http://dx.doi.org/10.1126/science.aad8828]

[29]   Lusher AL, McHugh M, Thompson RC. Occurrence of microplastics in the gastrointestinal tract of pelagic and demersal fish from the English Channel. Mar Pollut Bull 2013; 67(1-2): 94-9.
[http://dx.doi.org/10.1016/j.marpolbul.2012.11.028] [PMID: 23273934]

[30]   Greven AC, Merk T, Karagöz F, *et al.* Polycarbonate and polystyrene nanoplastic particles act as stressors to the innate immune system of fathead minnow (*Pimephales promelas*). Environ Toxicol Chem 2016; 35(12): 3093-100.
[http://dx.doi.org/10.1002/etc.3501] [PMID: 27207313]

[31]   Talvitie J, Heinonen M, Pääkkönen JP, *et al.* Do wastewater treatment plants act as a potential point source of microplastics? Preliminary study in the coastal Gulf of Finland, Baltic Sea. Water Sci Technol 2015; 72(9): 1495-504.
[http://dx.doi.org/10.2166/wst.2015.360] [PMID: 26524440]

[32]   Wright SL, Kelly FJ. Plastic and human health: a micro issue? Environ Sci Technol 2017; 51(12): 6634-47.
[http://dx.doi.org/10.1021/acs.est.7b00423] [PMID: 28531345]

[33]   Rochman C, Hoh E, Kurobe T, Swee JT. Ingested plastic transfers hazardous chemicals to fish and induces hepatic stress. Sci nd 2013.
[http://dx.doi.org/10.1038/srep03263]

[34]   Jabeen K, Su L, Li J, *et al.* Microplastics and mesoplastics in fish from coastal and fresh waters of China. Environ Pollut 2017; 221: 141-9.
[http://dx.doi.org/10.1016/j.envpol.2016.11.055] [PMID: 27939629]

[35]   Silva-Cavalcanti JS, Silva JDB, França EJ, Araújo MCB, Gusmão F. Microplastics ingestion by a common tropical freshwater fishing resource. Environ Pollut 2017; 221: 218-26.
[http://dx.doi.org/10.1016/j.envpol.2016.11.068] [PMID: 27914860]

[36]   Phillips MB, Bonner TH. Occurrence and amount of microplastic ingested by fishes in watersheds of the Gulf of Mexico. Mar Pollut Bull 2015; 100(1): 264-9.
[http://dx.doi.org/10.1016/j.marpolbul.2015.08.041] [PMID: 26388444]

[37]   Peters CA, Bratton SP. Urbanization is a major influence on microplastic ingestion by sunfish in the Brazos River Basin, Central Texas, USA. Environ Pollut 2016; 210: 380-7.
[http://dx.doi.org/10.1016/j.envpol.2016.01.018] [PMID: 26807984]

[38]   Raza A. Microplastics in freshwater systems: a review on its accumulation and effects on fishes 2018.
[http://dx.doi.org/10.20944/preprints201810.0696.v1]

[39]   Yousuf A, Naseer M, Journal NN-TI. Isolation and identification of microplastic particles from agricultural soil and its detection by fluorescence microscope technique. ThinkindiaquarterlyOrg nd 2019.

[40]   Depledge MH, Galgani F, Panti C, Caliani I, Casini S, Fossi MC. Plastic litter in the sea. Mar Environ Res 2013; 92: 279-81.
[http://dx.doi.org/10.1016/j.marenvres.2013.10.002] [PMID: 24157269]

[41]   Nel HA, Dalu T, Wasserman RJ. Sinks and sources: Assessing microplastic abundance in river sediment and deposit feeders in an Austral temperate urban river system. Sci Total Environ 2018; 612: 950-6.

[http://dx.doi.org/10.1016/j.scitotenv.2017.08.298] [PMID: 28886547]

[42]   Prata JC. Microplastics in wastewater: State of the knowledge on sources, fate and solutions. Mar Pollut Bull 2018; 129(1): 262-5.
[http://dx.doi.org/10.1016/j.marpolbul.2018.02.046] [PMID: 29680547]

[43]   Alimi OS, Farner Budarz J, Hernandez LM, Tufenkji N. Microplastics and nanoplastics in aquatic environments: aggregation, deposition, and enhanced contaminant transport. Environ Sci Technol 2018; 52(4): 1704-24.
[http://dx.doi.org/10.1021/acs.est.7b05559] [PMID: 29265806]

[44]   Cai L, Hu L, Shi H, Ye J, Zhang Y, Kim H. Effects of inorganic ions and natural organic matter on the aggregation of nanoplastics. Chemosphere 2018; 197: 142-51.
[http://dx.doi.org/10.1016/j.chemosphere.2018.01.052] [PMID: 29348047]

[45]   Ebere EC, Wirnkor VA, Ngozi VE. Uptake of microplastics by plant: a reason to worry or to be happy? World Sci News 2019; 131: 256-67.

[46]   Cox KD, Covernton GA, Davies HL, Dower JF, Juanes F, Dudas SE. Human consumption of microplastics. Environ Sci Technol 2019; 53(12): 7068-74.
[http://dx.doi.org/10.1021/acs.est.9b01517] [PMID: 31184127]

[47]   Smith M, Love DC, Rochman CM, Neff RA. Microplastics in seafood and the implications for human health. Curr Environ Health Rep 2018; 5(3): 375-86.
[http://dx.doi.org/10.1007/s40572-018-0206-z] [PMID: 30116998]

[48]   Yang CZ, Yaniger SI, Jordan VC, Klein DJ, Bittner GD. Most plastic products release estrogenic chemicals: a potential health problem that can be solved. Environ Health Perspect 2011; 119(7): 989-96.
[http://dx.doi.org/10.1289/ehp.1003220] [PMID: 21367689]

[49]   Brandt-Rauf P, Long C, Kovvali G, Li Y, Monaco R, Marion MJ. Plastics and carcinogenesis: The example of vinyl chloride. J Carcinog 2012; 11(1): 5.
[http://dx.doi.org/10.4103/1477-3163.93700] [PMID: 22529741]

[50]   Hermabessiere L, Dehaut A, Paul-Pont I, *et al.* Occurrence and effects of plastic additives on marine environments and organisms: A review. Chemosphere 2017; 182: 781-93.
[http://dx.doi.org/10.1016/j.chemosphere.2017.05.096] [PMID: 28545000]

CHAPTER 6

Microplastic Pollution, A Threat to Human Health: A Case Study at Thoothukudi, South India

Sekar Selvam[1,*] and Perumal Muthukumar[1]

[1] *Department of Geology, V.O.Chidambaram College, Tuticorin-628008, Tamilnadu, India.*

Abstract: Microplastic pollution has become a serious problem that affects all marine and terrestrial environments worldwide. In this study, we investigated microplastics in the beach sediments and thus we collected 18 sediments from seven locations in Thoothukudi coastal area. Microplastics were separated and recognized using visual and micro-Fourier Transform Infrared spectroscopy (µFT-IR) studies. Microplastics' concentration ranges from high concentrations (up to 53 particles kg^{-1} d.w) in the dune areas to visibly lower ranges compared to beach sediments (up to 27 particles kg^{-1} d.w). The majority of microplastics identified in collected sediments were polyethylene (PE), polypropylene (PP), fiber(F), cellulose(CL) and nylon(NY) . The result of this study can provide valuable background information about microplastic pollution by using Atomic Force Microscopy (AFM) and the outcome of the results shows the presence of microplastics that pollute the marine environment in Thoothukudi coastal area and the human health risk in these areas.

Keywords: Aquatic ecosystem, Biota, Density extraction, Environment, Extraction, Electrostatic separation, FTIR, Filtration, Grab sampler, Identification, Microplastic, Magnetic extraction, Planktonic net, Pyr-GC-MS, QA/QC, Quantification, Raman spectroscopy, SEM, Sediment, Sample processing, Sampling, Terrestrial ecosystem.

INTRODUCTION

The amount of plastic production on a global scale increases every year. Since 1950, the world's plastic and polymer production has rapidly increased from 1.5 to 311 million tons in 2014 335 million tons in 2016 [1, 2]. Because of the increasing production of those polymers and their usage in low biodegradability, plastic contamination has turned into a serious environmental issue. A large quantity of plastics mainly end up in the marine environment and disturb the food chain of the marine ecosystem.

---

* **Corresponding author Sekar Selvam:** Department of Geology, V.O.Chidambaram College, Tuticorin-628008, Tamilnadu, India; E-mail: geoselvam10@gmail.com

Rahul Singh and Neeta Raj Sharma (Eds.)

It has been identified that nearly 4.8–12.7 million tons of plastic debris or plastic waste enter the marine environment and this amount will be increased rapidly by an order of magnitude before 2025. There are a few meanings of microplastics, for instance; Gregory (2009) characterized them as the barely visible particles that go through a 500 μm but are held by a 67 μm sieve [3], while Imhof *et al.* (2013) characterized particles less than 1 mm as microplastics [4]. These days, it is broadly acknowledged that plastic particles' size less than 5 mm are considered as microplastics (MSFD Technical Subgroup on Marine Litter 2013). Microplastics can enter the marine environment as primary or secondary pollution that contaminates the surrounding marine environment. Primary microplastics are polymers made in small scale, *e.g.*, cosmetics [5] and medicine [6] parts or raw materials utilized for plastic production [7, 8]. Secondary microplastics are the results of physical (mechanical) degradation of greater plastic fragments [9 - 12]. In recent years, plastic pollution has received an increasing amount of interest from researchers, politicians, and the public. Microplastics (<5 mm) are a particular concern as they are suspected to accumulate in the environment and aquatic life.

Plastic is a generic term for man-made polymers that are most often prepared by polymerization of monomers from oil or gas. When not made from oil and gas, the polymer can be manufactured from coal, natural gasses, cellulose or latex from trees. The molecular backbone of a plastic polymer is typically composed of hydrocarbons and other naturally occurring compounds. Other chemicals, additives, are also added to the polymer to provide desirable properties, such as plasticizers that are added to improve the malleability of certain polymers. We now live in "a plastic world" where almost everything surrounding us is made of plastic, and it is hard to imagine a world free of this material. Plastic production has increased dramatically worldwide over the last 60 years, and is still increasing, with the current production of around 300 million tons yearly.

MICROPLASTICS IN MARINE ENVIRONMENT

Microplastics can pollute the marine environment in two ways: as primary pollution and as secondary contamination. Primary microplastics are polymers that are used in microscale items in everyday life, such as cosmetic and medical components, or as raw materials in the creation of plastics. Physical (mechanical) breakdown and bigger plastic particles produce secondary microplastics. Microplastics have been shown to damage creatures at all trophic levels in the marine environment, including worms, fish, sea turtles, birds, and mammals. Microplastics are mistaken for food by many creatures, and they selectively feed on them instead of food. The debris in the stomach might restrict appropriate food intake generated by a sensation of fullness and lessen feelings of hunger, so reducing the desire to eat. As a result, the pace of growth might be reduced and reproductive capacity and the ability to avoid predators. Ingested plastic can cause fast fatality if the gastrointestinal system is entirely clogged or seriously injured by abrasions and ulcers. At concentration levels, plastic waste in the marine environment comprises a variety of hydrophobic contaminants and trace metals. Some of these chemicals are added to plastics during the manufacturing process, while others are absorbed from the seawater. Organic pollutants can interfere with normal hormone functioning in some marine animals, causing mutations and cancer. Hydrophobic organic pollutants have been demonstrated to have a stronger affinity for polymers such as polyethylene, polypropylene, and polyvinyl chloride than for natural sediments. The micro-scale reduction of plastic litter apparently improves their absorption capabilities and accelerates the passage of hazardous chemicals from plastics into organisms. Microplastics have been found all over the world, not just in densely populated areas, but also in remote areas and deposition zones. Microplastics' sources in the maritime environment have not been well investigated. Harbours and shipyards, fisheries, waste water treatment facilities, coastal tourism, urban runoff, and rivers are all likely sources of their contributions. The fate of these organisms in beach and bottom sediments is currently unknown. Although it has long been recognised that sediments may acquire microplastics, no obvious association between microplastic concentration and sediment grain size has been discovered, as has been the case with organic matter and some other pollutants [13]. Aggregation with organic materials, on the other hand, may play a key role in microplastic transport. Long water exchange also encourages the building of contaminants in the area. The authors claim that this is the first research to evaluate the shape, concentration, and fate of microplastics in bottom and beach sediments. The primary sources of microplastics were also identified with their effect on human health.

The world's seas and oceans are subjected to different kinds of threats, of which the accumulation of anthropogenic debris is a major and worldwide problem that has been an environmental concern for decades. Despite increased worldwide attention, the accumulation of these materials in the environment is seen as an issue due to rising global plastic manufacturing (280 million tonnes in 2011) and continued poor plastic waste disposal. It has been established in the previous decade that big plastic items in the marine environment break up into smaller particles with dimensions as thin as a safe micrometre, known as microplastics. Moreover, additional sources of microplastics have been identified. Microplastic particles present in cosmetics and those fibres from fabrics such as polyester and polyamide present in wastewater used domestically are not retained during the treatment of sewage and can thus enter the coastal environment. Many authors have defined microplastics as particles smaller than 5 mm, while others have set the upper size limit at 1mm. While the value of 5 mm is more commonly used, 1 mm is a more intuitive value (*i.e.* 'micro' refers to the micro metre range).

Our marine environment is subjected to many different kinds of threats. One of these threats is marine debris. The accumulation of anthropogenic debris in the marine environment is an increasing problem worldwide. This debris is not only aesthetically displeasing, but can also be a nuisance to boaters and the shipping industry, and can adversely affect marine biota [14]. Complications in nets, fishing line, ropes, and other debris, which can impose cuts and wounds or cause suffocation or drowning; and ingestion, which can cause blocks in throats or digestive tracts, are among the reported consequences on marine creatures. Some animals may even starve to death because a material that does not pass through the stomach might fool them into thinking they are no longer hungry, prompting them to quit consuming more. Beach visitors can be damaged by broken glass, medical waste, dropped fishing lines, and syringes, which are all examples of marine litter [15].

MATERIALS AND METHODS

Study Site

Thoothukudi coast is the southernmost part of the coast in the Tamilnadu state, India. It is located between the latitude 8°8' to 8°40'N and longitudes 77°35' to 78°8'E lying in Thooothukudi district of Tamilnadu. The Coastal zone considered for the present study covers 1428sq.km and falls under a semiarid climatic zone. The mean maximum temperature of this region is 36.5°C(May) and minimum temperature is 22.3°C(Dec).The average annual rainfall is only 760mm, in which the contribution of NE Monsoon (Oct-Dec) seems to be more than the SW Monsoon (Jul-Sept). In this region, frequent failure of NE moon results in the depletion of freshwater resources. The region is topographically low and flat. A teri soil mound is located in the central part of this area. Thoothukudi is covered by the geological formations of Archaean, Tertiary and Recent to Subrecent periods. In this area, Tiruchendur is one of the famous pilgrimage centres in south India and nowadays, it is one of the tourist spots.

Tuticorin is one of the coastal cities in Tamilnadu that have experienced rapid growth in the last few decades. People from the nearby villages migrated to Tuticorin for their rapid growth of population and unplanned growth of the city both horizontally in all directions. Due to the enhanced importance of the city and its environment, people migrate towards the city, and they occupy lands for businesses, commercial and residential purposes. Many of the developments that have come up in the recent years have affected the study area in a drastic way. So there is a need for proper planning for the careful handling of this alarming situation. The dumping of the residential solid waste and the untreated domestic wastewater that gets mixed with the sea water, affect the coastal ecology. In view of the above, it becomes imperative to study the macroplastics aspects of pollution of Tuticorin.

Sampling on the Shorelines

Sediment samples was collected on the southern shore of Thoothukudi to investigate the influence of catchment management on the quality and quantity of microplastics on beaches (Fig. **1**).We sampled and collected sediments from seven sites along the shoreline. Nearly 18 sediment samples are collected along the shoreline. The sample collection in coastal beaches is easier and more convenient. Samples were collected using a manta trawl, a rectangular opening and sediments are collected in a polyethylene sample bags. There is not one specific way to be followed during sampling process. It depends on individual and the place of sampling. Most researches prefer to use tidelines, sampling depth to collect samples from beaches [16]. The procedure of sample collection is simple as it is done by using stainless steel tools like a shovel or spoon. Using latex gloves and cotton clothes during the sampling process helps to minimize the contamination in samples. Sediment sample collection process was done in following way: First try to find the high tide line, where the debris that washed away with tide get accumulated. Then randomly select the location tide line and place the quadrate with the tide line. After selecting the location make sure to remove large pieces of debris from the selected location. At last use shovel or stainless-steel spoon to pick up the sample from the top 3 cm of sand. Then the collected sample should be placed in sample collection bottles and marked with the label has been information on date of sample collection, place, GPS location.

The sampling sites were selected based on the availability of the debris and length of the high tide and berm region. The present study area primary source of marine debris are from household materials and daily usage items including water/cool drinks bottles, use and throw polyethylene covers like milk covers and broken plastic materials. This area is one of the tourist spot and peoples hereby visit from various places may use many plastic bottles and polyethylene bags and they end up in coastal areas.

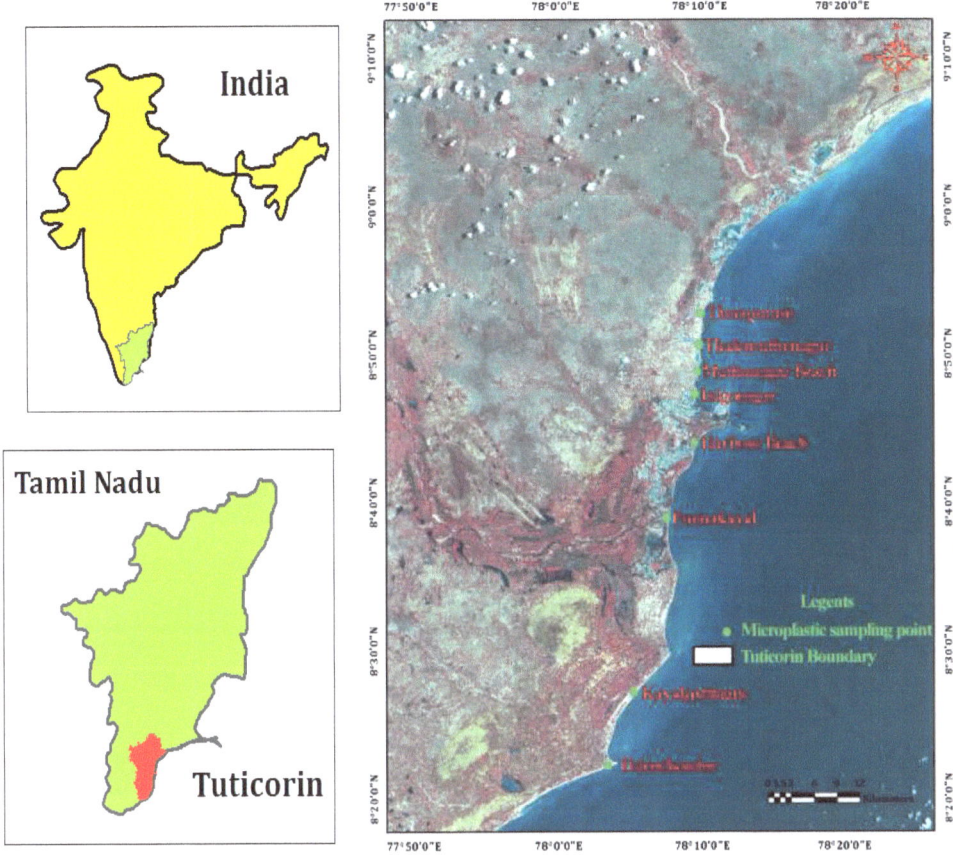

Fig. (1). Sampling locations of the Study area.

SAMPLE SEPARATION AND PURIFICATION

Flotation Process

Sample separation and purification were done in many ways. For microplastics sampling, we used the density separation method. It is used to separate low-density particles from higher-density particles such as sand, mud, and sediments. Microplastics, such as PP and PE, have lower density in the comparison to sea water (1.10 g/cm3) [16]. In the case of higher densities, microplastics like PVC (1.40 g/cm3) or greater than that, different kinds of density solutions were used to separate microplastics from them. In the case of samples with higher density, (PVC) may not be completely extracted during the process.

The process of density separation begins with mixing a salt solution with a sample and shaking it properly to homogenize the slurry [16]. Then it is allowing to settle down for few hours, which let higher densities particles (sand) to settle down on the base. Filtration was used to separate the solution above the sediment, and the microplastics were extracted using the flotation process.

Sample Purification

Sample purification was done to obtain reliable data about microplastics present in samples. The surface of microplastic samples collected from various sections has a lot of organic materials adhering onto it. To achieve so, a 30 percent H_2O_2 solution has been commonly utilized [16]. Many other digestion techniques are also employed to take the organic matter away from the sample. A recent study suggested that the presence of vegetal materials like algae, seagrasses along with various small residues is abundant in microplastics samples obtained from beaches.

IDENTIFICATION PROCESS

After sample preparation, microplastics were identified from various techniques. Among all the microplastics identification techniques, visual identification is the most common approach to identify microplastics. This is then followed by confirmation through the chemical composition by using optical and spectroscopic techniques.

Visual Identification

This is the first step toward the microplastics identification process. In case of large microplastics, it was sorted out directly. While in the case of small sized microplastics, there is a need for further observation under a microscope. Because the risks of misidentification are quite significant, visual identification is not relevant to particles smaller than 500 m [17].The drawback of this method is the size limitation of the sample, for instance: particles below a certain size cannot be differentiated.

Identification by FTIR Spectroscopy

This method was suitable for determining the polymer origin and composition of microplastics in samples [17]. The main advantage of using this technique is that it excites the molecular vibrations while interacting with the sample. In case of a plastic sample, it makes it easier with this technique to obtain IR spectra having discrete band patterns that are extremely distinctive. FTIR spectroscopy also offers information on the weathering of tested plastic particles based on the oxidation level detected [17]. To obtain high quality data and to reduce the measurement time, a resolution of 8 cm^{-1} is suggested (Bergmann *et al.*, 2015). The main drawback of this technique is the size limitation of sample. There is still problem to analyze particles size < 1 μm Fig. (**2a, b**).

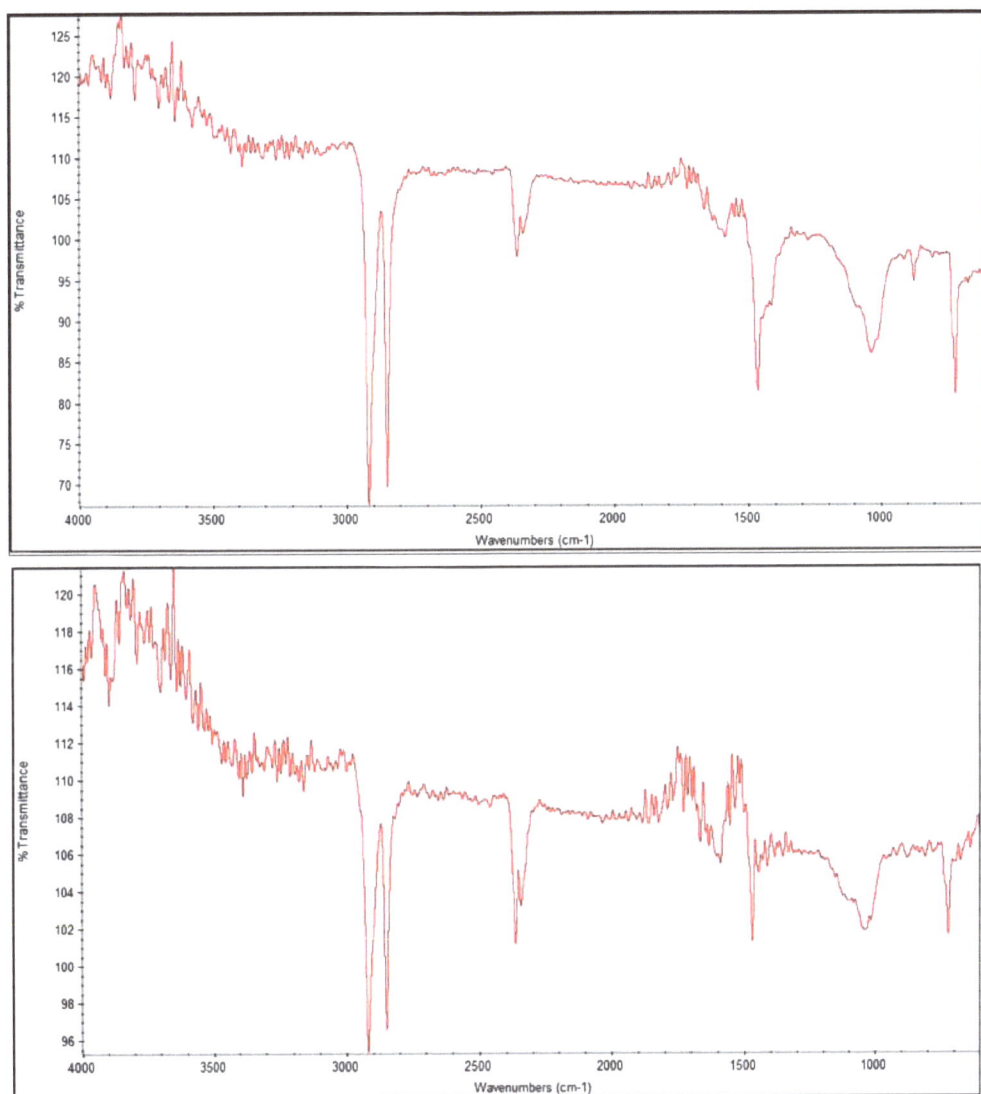

Fig. (2a). μ-FTIR peaks of the micro plastic particles from Thoothukudi beach T1-Polethylene T2-Nylon.

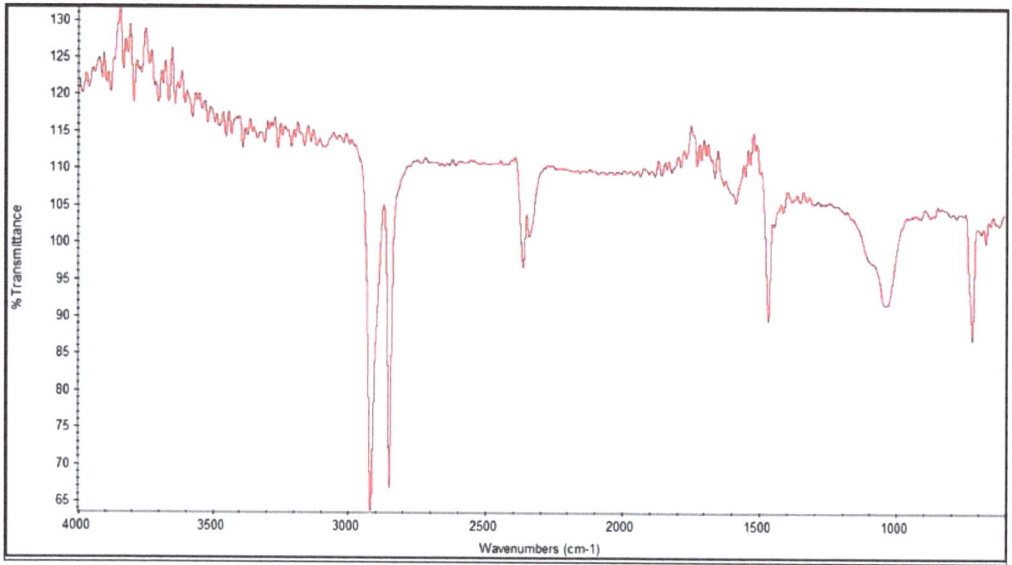

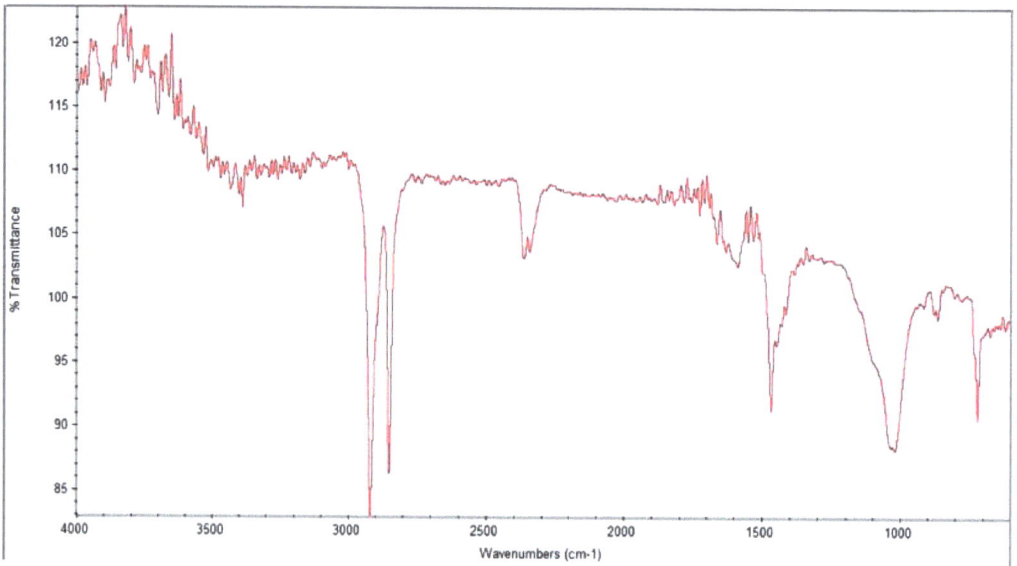

Fig. (2b). μ-FTIR peaks of the micro plastic particles from Thoothukudi beach T3-Fiber T4-Cellulose.

Identification of Surface Morphology by AFM

The surface elemental composition was determined in Atomic force microscopy (AFM). AFM has become a standard technique, both to visualize surface topography and microstructure as well as to probe materials properties. The advantages of AFM compared to, for example, electron microscopy are the ease of sample preparation and the possibility not only to visualize surface microstructure and topography, but also to probe a broad range of materials' properties (Fig. **3**).

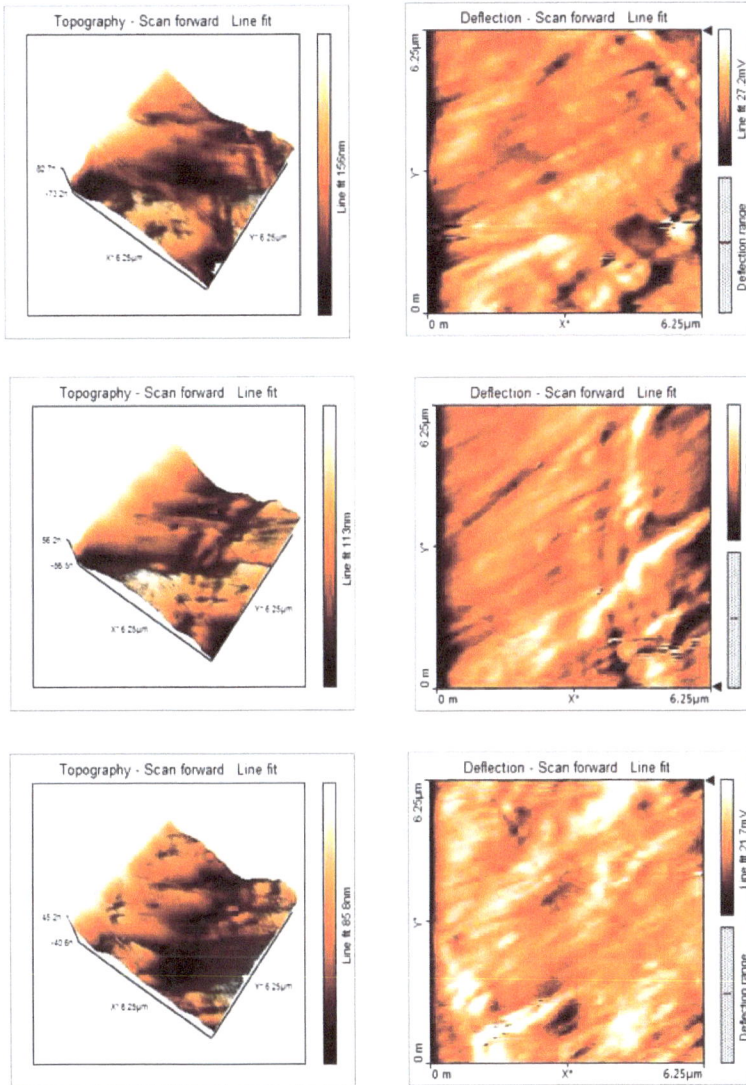

Fig. (3). Surface elemental compositions of Nylon, Polyethylene and Polyester microplastics in Thoothukudi beach.

RESULTS AND DISCUSSION

The morphology of the debris, especially microplastic is a proxy tool to identify their potential origins. Fishing lines, clothes, and other textiles are common sources of line/fiber materials, while bags and wrapping materials are the most prevalent source of film materials. The polyethylene water bottles were found everywhere on the Thoothukudi coastal area. The marine debris also includes micro plastic particles used in cosmetic materials, resin pellets peelings [18], household items such as insect repellents, sunscreen and *etc* [19, 20], others include synthetic clothing [21]. The nylon material is primarily contributing to the coast from fishing nets, nylon ropes. The polystyrene materials were transported into the coastal environment from nearby urban areas and fishing hamlets. The polystyrene materials are commonly used as packaging materials and floaters during fishing practices. The water bottles and polystyrene pieces were also found in most parts of the coastal berm area, probably due to the action of wind. However, the majority of the marine debris are dumped in the high tide region due to high energy wave's action.

Thoothukudi is one of the Industrial cities in south Tamil Nadu, India. In the present study location, all sediment samples contained large number of microplastic particles in pipetted and decanted solutions. In pipetted solutions, particles ranged from 10 particles to 100 particles per sample with typical values between 20 to 60 particles and some are easily visible to identify. In the decanted solutions, several 100 to 1000 particles were extracted onto the zooplankton net filters. The vast majority of these particles was visually indistinct from natural sediment grains. Rinsing with the deionised water caused about half of the extracted particles to sink immediately, while the remainder floated on the surface initially, but sank upon contact with the lancet. The majority of extracted particles are therefore indistinguishable in density and in appearance from natural sediment [22].

As expected from the artificially enriched samples, coloured particles and fibres stand out most prominently in the sediment samples. Both particles and fibres are dominated by intense blue, violet, green, or red colours, providing strong evidence for their anthropogenic origin. Without spectroscopic confirmation, natural anthropogenic coloured fibres cannot be distinguished from synthetic fibres here. However, long persistence times are expected for synthetic materials, while natural fibres such as cotton or wool can be expected to dissolve under marine conditions more rapidly. Coloured particles and fibres will therefore be used as the dominant tracers for microplastic contaminations in the following discussion. Almost all of the samples included coloured fibres, and 5 of the 7 sediment samples had brightly coloured particles. In the tourist area, only outer beach contained coloured particles in the three other locations of the coast line.

Various other types of microplastics were identified, including nylon (NY) (18.9%),cellulose(CL) (8.9%), polyethylene (PE) (33.9%). polypropylene (PP) (18.9%) and some microplastics or non-plastic particles, were also unidentified through u-FTIR. The coastal debris was dominated by polyethylene bottles and fibrous fishnet materials. The distribution of the Coastal debris along the coastal region is chiefly controlled by winnowing action of sea and waves and aeloian action along the berm region. The main source of microplastic pollution and contamination in Thoothukudi area is tourist's pollution. The polyethylene bags that are mainly contributed to the coast from river channels to sea, the usage of plastic bottles and polyethylene bags by tourists the nylon material also contributed to the coast from fishing nets and nylon ropes.

MICROPLASTICS AS A HUMAN THREAT

Microplastics are considered as an emerging environmental contaminant [23] and so there is a dearth of convincing evidence on adverse human health effects. The Mps articles did not demonstrate any direct evidence of human health risk due to exposure to MPs. Fortunately, there is a growing interest among scientists and researchers in exploring the knowledge gaps.

Microplastics in beach sediments is one of the main pathway to the food chain. Where coastal sediments can affect the coastal ecosystem. Microplastics are consumed by a variety of terrestrial and marine animals, including molluscs, fish [24], shrimp and crabs and may transit through food chains to humans. As a result, it would pose severe threats to both the biota and human health. MPs are made up of a range of polymers that are generated from a chain of monomers through a polymerization process in which unreacted monomers and potentially hazardous additives are present. As such, the selection of biodegradable and non-toxic monomers is of paramount importance toward the development of environmentally friendly biodegradable polymer materials. Chemical components in more than 50 percent of plastics are harmful, according to a hazard-ranking model based on the United Nations' Globally Harmonized System of categorization and labelling of chemicals. Many parameters are used to estimate the ecological risk of MPs in terrestrial and marine sediments.

Consumption of seafood is one way for humans to be exposed to microplastics [25]. Global seafood consumption accounted for 6.7 percent of total protein consumption and around 17 percent of animal protein consumption in 2018. Microplastics can be consumed by a wide range of marine species due to their microscopic size. Direct or indirect trophic transfer ingestion is possible (*e.g.*, up the food web). Microplastics have been found in planktonic creatures and larvae at the bottom of the food chain [18], small and big invertebrates [26], and fish [27]. Because plastic particles are commonly concentrated in an organism's digestive tracts, bivalves and tiny fish that are swallowed whole are more likely to expose humans to microplastics.

The type of the hazardous chemical, exposure parameters, individual sensitivity, and hazard controls all influence the intensity of adverse effects caused by exposures. Although the physical effects of accumulated microplastics are less well understood than the distribution and storage of toxicants in the human body, preliminary research has shown a number of potentially harmful effects, including increased inflammatory response, size-related toxicity of plastic particles, chemical transfer of adsorbed chemical pollutants, and disruption of the gut microbiome [28].

Direct exposure to POPs and other chemicals related to microplastics can have a negative impact on biological systems and offer a particular risk to children and animals, even at low doses (GESAMP; ATDSR 2015). To assess risk at lower exposure levels or perform low-dose extrapolations, current standards for toxicity testing of chemical components employ high contaminant concentrations from a single compound. Concerns about low-dose pollutants or mixed groupings of contaminants are not captured by this technique. This strategy also makes it difficult to account for non-linear exposure relationships. As a result, current methodologies are unable to produce data that accurately reflects the potential hazard caused by microplastic-associated compounds. Microplastics have the potential to damage individuals in both physical and chemical ways. Microplastics are abundant in the marine environment and are degrading marine organisms at a growing rate. The excretory system of the human body removes microplastics, with faeces capable of disposing of >90 percent of ingested microplastics. The intensity of adverse effects from exposure is determined by the hazardous chemical's composition, exposure parameters, individual vulnerability, and risk management strategies.

Biota and microplastics frequently interact through ingestion. Microplastics and their related compounds have different fates and impacts in different species and settings [29]. The mixture of microplastics and related compounds has been shown to enhance toxicity in laboratory experiments [30]. However, determining if toxicological effects on people are present is challenging [31]. Chemicals emitted by microplastics are thought to be little in comparison to those emitted by other food components in animals [29]. Microplastics and their components can cause localised particle toxicity, but chronic exposure with a cumulative impact is more dangerous. In conclusion, more research is needed to quantify the chemical dosage to people resulting from microplastics in seafood, as well as the impacts of these chemicals, including investigations of seafood intake, chemical characterisation in seafood, and kinetic studies.

Microplastics and their components can cause localised particle toxicity, but chronic exposure with a cumulative impact is more dangerous. To fill these research gaps, scientists should assess the relative effect of microplastics as an exposure pathway. Identifying sorbed contaminant bioavailability and using biomonitoring technologies to establish acceptable toxicological exposure levels for chronic exposure to microplastics and their components would also be beneficial [32]. Industries play a key role in the production of microplastics across the supply chain in the form of primary microplastics utilized in industrial processes and secondary microplastics in the study area. These microplastics are also harmful to people's health.

EFFECTS OF MICROPLASTICS

Plastic waste in the ocean is clearly a worldwide problem that affects all major water bodies above and below the surface. They have the potential to harm individuals, wildlife, and coastal communities' economic health [15].

Humans

Sewage related plastic debris, medical waste and other potential biohazards are considered as a potential danger to human health, either when stranded on beaches or when circulating in coastal waters. Marine plastic debris can lead to problematic entanglement, and particularly the entanglement of divers in monofilament gill nets which are floating around in the water column or attached to wrecks.

Coastal Communities

Damage to coastal communities caused by marine litter can be grouped into a number of general categories. These include damage to fisheries, fishing boats and gear, damage to cooling water intakes in power stations, contamination of beaches (requiring cleaning operations), contamination of commercial harbours and marinas (demanding cleaning operations), and contamination of coastal grazing land, causing injury to livestock (UNEP, 2005). Fishermen reported problems with propeller fouling, blocked intake pipes and damaged drive shafts. Marine litter-related damage includes also safety risks at sea (demanding rescue services) due to fouling of propellers [33]. The dumpage of waste materials can produce marinedebris that makes shorelines unattractive and potentially hazardous which forces the community peoples and government to allocate favourable amount of funds for the maintenance of coastal beaches. The presence of marine debris are mainly discouraged the common peoples to visit coastal areas.

Wildlife

Marine wildlife is possibly the group of organisms mostly affected by the debris. The threats to marine life are primarily mechanical due to ingestion of plastic debris and entanglement in packaging bands, synthetic ropes and lines, or drift nets [34, 35]. Many birds have been found to contain small items of debris in their stomachs. This is the result of their mistaking the litter for food (*e.g* [34, 36].). Some seabirds have been shown to pick certain plastic forms and colours, mistaking them for possible prey [14]. The same could be true for fish, as various species were found to have plastic debris in their guts which were all of the same type of white plastic spherules, indicating selective feeding [37]. The consequences of this ingestion of debris include reduction of food uptake due to the plastics reducing the storage volume of the stomach and the feeding stimulus. This can result in a loss of general fitness, as well as a blockage of stomach enzyme secretion, reduced steroid hormone levels, delayed ovulation, and reproductive failure [14]. It is also suggested that plastic pellets could be a route for PCBs (or other contaminants) into marine food chains (*e.g* [37].). There is limited evidence for this bio-transfer: Ryan *et al.* (1988) related PCBs in bird tissues to that associated with ingested plastic particles [38]. Besides fish and birds, sea turtles, whales and dolphins have also been found to ingest plastic, sometimes with death as a result [14].

Another severe hazard to marine species is entanglement in marine debris, particularly discarded fishing gear. Once an animal is entangled, it may drown, have its ability to catch food or to avoid predators impaired, or incur wounds from abrasive or cutting action of attached debris. Sea mammals in particular, are vulnerable to entanglement [14].

MITIGATION

The status of evidence relating microplastics to possible human health risk has been presented in the above microplastic hazards to human health section. Microplastics, chemical toxicity, and chronic exposure to microplastics may constitute a health concern to humans, particularly when direct exposure to plastic and localised chemicals increases. While there are still some gaps, complementary sets of research point to possible exposures and risks from both particles and related substances. The influence of microplastics on human health is unknown, but it cannot be overlooked, and it provides one argument for reducing the amount of plastic entering the environment. Governments, business, and civil society all must play critical roles.

CONCLUSION

The study focused on investigating the presence of plastic pollution in the marine environment in Thoothukudi coastal area and this is the preliminary study in this area. The obtained data indicate that the degree of urbanisation in the closest region of catchment may be a significant factor determining the microplastics content in the beach sediments of Thoothukudi, Tamil Nadu's southern coast. Microplastics were present in high concentrations (up to 53 particles kg^{-1} d.w) in the dune areas and visibly lower compared to beach sediments (up to 27 particles kg^{-1} d.w) near continental shelf regions. Variable impacts of common debris on the health of marine wildlife were vastly identified, with complication by fishing-related gear and plastic bags emerging as the greatest threat to sea turtles and seabirds as marine mammals. Standardized data collection methods are used for microplastic occurrence in the environment and food stuffs, followed by exposure assessment for dietary intake. By collecting data on the presence and quantity of degraded plastics in food, and data on the translocation of microplastics through the aquatic food web and human food system, methods are developed for assessing physical and chemical changes of micro- and nanoplastics when interacting with biological systems. Toxicity exposure data is analysed to evaluate mixtures of various additives. However a wide variety of other items posed at least some threat to these organisms through either ingestion, contamination or both, suggest that a comprehensive approach to preventing plastics from entering the ocean is vitally needed. While there is still much to learn, closing these gaps is critical to achieving the dual aim of increasing seafood consumption while also safeguarding customers from the harmful health consequences of microplastics in the marine ecosystem. This has an impact on qualitative study findings, which offer the foundation for conclusions concerning their origins in beach sediments.

ACKNOWLEDGEMENTS

The first author Sekar Selvam sincerely acknowledges the Department of Science and Technology -SERB-ECR_ Government of India, New Delhi (Grant No: F:ECR/2018/001749) for financial assistance. The authors would also like to express their gratitude to Shri A.P.C.V.Chockalingam, Secretary, and Dr.C.Veerabahu, Principal, V.O.C.College, Tuticorin, for their assistance and advice.

REFERENCES

[1]   Comăniță ED, Hlihor RM, Ghinea C, Gavrilescu M. Occurrence of plastic waste in the environment: ecological and health risks. Environ Eng Manag J 2016; 15(3): 675-85.
[http://dx.doi.org/10.30638/eemj.2016.073]

[2]   Kaliszewicz A, Winczek M, Karaban K, *et al.* The contamination of inland waters by microplastic fibres under different anthropogenic pressure: Preliminary study in Central Europe (Poland). Waste Manag Res 2020; 38(11): 1231-8.
[http://dx.doi.org/10.1177/0734242X20938448] [PMID: 32659207]

[3]   Gregory MR. Environmental implications of plastic debris in marine settings—entanglement, ingestion, smothering, hangers-on, hitch-hiking and alien invasions. Philos Trans R Soc Lond B Biol Sci 2009; 364(1526): 2013-25.
[http://dx.doi.org/10.1098/rstb.2008.0265] [PMID: 19528053]

[4]   Imhof HK, Ivleva NP, Schmid J, Niessner R, Laforsch C. Contamination of beach sediments of a subalpine lake with microplastic particles. Curr Biol 2013; 23(19): R867-8.
[http://dx.doi.org/10.1016/j.cub.2013.09.001] [PMID: 24112978]

[5]   Zitko V, Hanlon M. Another source of pollution by plastics: Skin cleaners with plastic scrubbers. Mar Pollut Bull 1991; 22(1): 41-2.
[http://dx.doi.org/10.1016/0025-326X(91)90444-W]

[6]   Patel MM, Goyal BR, Bhadada SV, Bhatt JS, Amin AF. Getting into the Brain. CNS Drugs 2009; 23(1): 35-58.
[http://dx.doi.org/10.2165/0023210-200923010-00003] [PMID: 19062774]

[7]   Turner A, Holmes LA. Adsorption of trace metals by microplastic pellets in fresh water. Environ Chem 2015; 12(5): 600-10.
[http://dx.doi.org/10.1071/EN14143]

[8]   Naqash N, Prakash S, Kapoor D, Singh R. Interaction of freshwater microplastics with biota and heavy metals: a review. Environ Chem Lett 2020; 18(6): 1813-24.
[http://dx.doi.org/10.1007/s10311-020-01044-3]

[9]   Zbyszewski M, Corcoran PL, Hockin A. Comparison of the distribution and degradation of plastic debris along shorelines of the Great Lakes, North America. J Great Lakes Res 2014; 40(2): 288-99.
[http://dx.doi.org/10.1016/j.jglr.2014.02.012]

[10]   Galgani F, Hanke G, Maes T. Global distribution, composition and abundance of marine litter.InMarine anthropogenic litter. Cham: Springer 2015; pp. 29-56.

[11]   Koelmans AA, Besseling E, Shim WJ. Nanoplastics in the aquatic environment. Critical review. Marine anthropogenic litter 2015; 325-40.

[12]   Manzoor S, Kaur H, Singh R. Existence of microplastic as pollutant in harike wetland: an analysis of plastic composition and first report on ramsar wetland of india. Current World Environment 2021; 16(1).
[http://dx.doi.org/10.12944/CWE.16.1.12]

[13]   Manzoor S, Naqash N, Rashid G, Singh R. Plastic material degradation and formation of microplastic in the environment: A review. Mater Today Proc 2021.

[14]   Derraik JGB. The pollution of the marine environment by plastic debris: a review. Mar Pollut Bull 2002; 44(9): 842-52.
[http://dx.doi.org/10.1016/S0025-326X(02)00220-5] [PMID: 12405208]

[15]   Sheavly SB, Register KM. Marine debris & plastics: environmental concerns, sources, impacts and solutions. J Polym Environ 2007; 15(4): 301-5.
[http://dx.doi.org/10.1007/s10924-007-0074-3]

[16]   Mai L, Bao LJ, Shi L, Wong CS, Zeng EY. A review of methods for measuring microplastics in

aquatic environments. Environ Sci Pollut Res Int 2018; 25(12): 11319-32.
[http://dx.doi.org/10.1007/s11356-018-1692-0] [PMID: 29536421]

[17]   Bergmann M, Gutow L, Klages M. Marine anthropogenic litter. Springer Nature 2015.
[http://dx.doi.org/10.1007/978-3-319-16510-3]

[18]   Cole M, Lindeque P, Halsband C, Galloway TS. Microplastics as contaminants in the marine environment: A review. Mar Pollut Bull 2011; 62(12): 2588-97.
[http://dx.doi.org/10.1016/j.marpolbul.2011.09.025] [PMID: 22001295]

[19]   Castañeda RA, Avlijas S, Simard MA, Ricciardi A. Microplastic pollution in St. Lawrence River sediments. Can J Fish Aquat Sci 2014; 71(12): 1767-71.
[http://dx.doi.org/10.1139/cjfas-2014-0281]

[20]   Duis K, Coors A. Microplastics in the aquatic and terrestrial environment: sources (with a specific focus on personal care products), fate and effects. Environ Sci Eur 2016; 28(1): 2.
[http://dx.doi.org/10.1186/s12302-015-0069-y] [PMID: 27752437]

[21]   Alomar C, Estarellas F, Deudero S. Microplastics in the Mediterranean Sea: Deposition in coastal shallow sediments, spatial variation and preferential grain size. Mar Environ Res 2016; 115: 1-10.
[http://dx.doi.org/10.1016/j.marenvres.2016.01.005] [PMID: 26803229]

[22]   Venkatramanan S, Chung SY, Selvam S, *et al.* Characteristics of microplastics in the beach sediments of Marina tourist beach, Chennai, India. Mar Pollut Bull 2022; 176: 113409.
[http://dx.doi.org/10.1016/j.marpolbul.2022.113409] [PMID: 35168068]

[23]   Bhuyan MS, Venkatramanan S, Selvam S, *et al.* Plastics in marine ecosystem: A review of their sources and pollution conduits. Reg Stud Mar Sci 2021; 41: 101539.
[http://dx.doi.org/10.1016/j.rsma.2020.101539]

[24]   Selvam S, Jesuraja K, Venkatramanan S, Roy PD, Jeyanthi Kumari V. Hazardous microplastic characteristics and its role as a vector of heavy metal in groundwater and surface water of coastal south India. J Hazard Mater 2021; 402: 123786.
[http://dx.doi.org/10.1016/j.jhazmat.2020.123786] [PMID: 33254795]

[25]   Selvam S, Manisha A, Venkatramanan S, Chung SY, Paramasivam CR, Singaraja C. Microplastic presence in commercial marine sea salts: A baseline study along Tuticorin Coastal salt pan stations, Gulf of Mannar, South India. Mar Pollut Bull 2020; 150: 110675.
[http://dx.doi.org/10.1016/j.marpolbul.2019.110675] [PMID: 31669711]

[26]   Murray F, Cowie PR. Plastic contamination in the decapod crustacean Nephrops norvegicus (Linnaeus, 1758). Mar Pollut Bull 2011; 62(6): 1207-17.
[http://dx.doi.org/10.1016/j.marpolbul.2011.03.032] [PMID: 21497854]

[27]   Selvam S, Manisha A, Roy PD, *et al.* Microplastics and trace metals in fish species of the Gulf of Mannar (Indian Ocean) and evaluation of human health. Environ Pollut 2021; 291: 118089.
[http://dx.doi.org/10.1016/j.envpol.2021.118089] [PMID: 34536648]

[28]   Wright SL, Kelly FJ. Plastic and human health: a micro issue? Environ Sci Technol 2017; 51(12): 6634-47.
[http://dx.doi.org/10.1021/acs.est.7b00423] [PMID: 28531345]

[29]   Lusher A, Hollman P, Mendoza-Hill J. Microplastics in fisheries and aquaculture: status of knowledge on their occurrence and implications for aquatic organisms and food safety. FAO 2017.

[30]   Browne MA, Niven SJ, Galloway TS, Rowland SJ, Thompson RC. Microplastic moves pollutants and additives to worms, reducing functions linked to health and biodiversity. Curr Biol 2013; 23(23): 2388-92.
[http://dx.doi.org/10.1016/j.cub.2013.10.012] [PMID: 24309271]

[31]   Talsness CE. Overview of toxicological aspects of polybrominated diphenyl ethers: A flame-retardant additive in several consumer products. Environ Res 2008; 108(2): 158-67.
[http://dx.doi.org/10.1016/j.envres.2008.08.008] [PMID: 18949835]

[32]   Welle F, Franz R. Microplastic in bottled natural mineral water – literature review and considerations on exposure and risk assessment. Food Addit Contam Part A Chem Anal Control Expo Risk Assess 2018; 35(12): 2482-92.
[http://dx.doi.org/10.1080/19440049.2018.1543957] [PMID: 30451587]

[33]   Hall AE. Crop responses to environment.Environment & Agriculture. 1st Edition. Boca Raton: CRC press 2000; p. 248.
[http://dx.doi.org/10.1201/9781420041088]

[34]   Laist DW. Impacts of marine debris: entanglement of marine life in marine debris including a comprehensive list of species with entanglement and ingestion records. InMarine debris.Marine Debris. New York, NY: Springer 1997; pp. 99-139.
[http://dx.doi.org/10.1007/978-1-4613-8486-1_10]

[35]   Quayle DV. Plastics in the marine environment: problems and solutions. Chem Ecol 1992; 6(1-4): 69-78.
[http://dx.doi.org/10.1080/02757549208035263]

[36]   Day RH, Wehle DH, Coleman FC. Ingestion of plastic pollutants by marine birds. inproceedings of the workshop on the fate and impact of marine debris. US Dept.. Commerce. 1985; pp. 2: 344-86.

[37]   Venrick EL, Backman TW, Bartram WC, Platt CJ, Thornhill MS, Yates RE. Man-made objects on the surface of the central North Pacific Ocean. Nature 1973; 241(5387): 271.
[http://dx.doi.org/10.1038/241271a0] [PMID: 4701885]

[38]   Ryan PG. The characteristics and distribution of plastic particles at the sea-surface off the southwestern Cape Province, South Africa. Mar Environ Res 1988; 25(4): 249-73.
[http://dx.doi.org/10.1016/0141-1136(88)90015-3]

CHAPTER 7

Microplastic as a Multiple Stressor

Savita Bhardwaj[1], **Dhriti Sharma**[1], **Tunisha Verma**[1] and **Dhriti Kapoor**[1,*]

[1] *Department of Botany, School of Bioengineering and Biosciences, Lovely Professional University, Phagwara (Punjab), India*

Abstract: The presence of microplastics (MPs) throughout the world causes a serious threat to the functionality and vigor of the ecosystem, which is present in almost all habitats, such as in aquatic, atmospheric and terrestrial habitats, and is also found in human consumables. Recently it has been found that MPs have entered the human body through the food chain from terrestrial agriculture. Migration and retention of MPs in the soil are controlled by the interaction between MPs and various environmental factors. There is an immense need in real-world environments to understand the migration properties and key mechanisms of MPs. Various organisms such as plants, animals, different microorganisms present in the soil, *etc.* are impacted by the presence of toxic MPs in the environment. Therefore, to ensure food safety and sustainable agriculture, MPs should be treated as a future threat and attention should be given to understand the mechanisms of transport and ecotoxicological effects of contaminants released from MPs. The aim of the present chapter is to emphasize the impact of MPs on various organisms present in the ecosystem and their interaction with other contaminants.

Keywords: Antioxidative enzymes, Contaminants, Ecosystem, Embryonic development, Fecundity, Food chain, Hazard, Human health, Microbes, Mortality, Microplastics, Multiple stressors, Nanoplastics, Non-degradation, Oxidative stress, Plant growth, Reactive oxygen species, Seed germination, Stress responsive genes, Toxicity.

INTRODUCTION

Plastic has become a ubiquitous material nowadays owing to its properties such as low weight, durability, rust resistance along with low electrical and thermal conductivity [1 - 3]. Though its mass production, mismanaged usage, and non-degradation coupled with inappropriate waste disposal, have made it a serious ecological hazard whereby it causes acute damage to the environment [4, 5].

* **Corresponding author Dhriti Kapoor:** Department of Botany, School of Bioengineering and Biosciences, Lovely Professional University, Phagwara (Punjab), India; E-mail: dhriti405@gmail.com

Rahul Singh and Neeta Raj Sharma (Eds.)
All rights reserved-© 2023 Bentham Science Publishers

Further, the widespread applications of plastics have resulted in releasing them on a large scale into aquatic and terrestrial environments where they will persist for centuries owing to the difficulty in their natural degradation. Plastic is being extensively deposited in the world's oceans as 70–80% of plastics are transported by rivers [6]. Natural ecosystems contain plastic debris of almost every shape and size and until recently, the maximum damage is caused by large-sized plastic articles. For example, several problems have been reported by different studies like mortality in sea turtles [7] and entangled seabirds following ingestion [8, 9].

However, microplastics appeared as a facet of great public concern in current times, and are defined as materials having dimension with an upper limit of <1 mm, as per the recommendations of Hartmann *et al.* [10]; whereas the size of plastic debris varies from microscopic particles to pieces to meters in dimensions. MPs exist contrary to the large-sized plastic refuse in the ecosystem as these are comparatively minute in size being almost invisible and cannot be eliminated from territories for reuse. MPs are considered ubiquitous anthropogenic contaminants in the ecosystem as they pose serious health effects and originate from different sources in the ecosystem.

Categorization of microplastics has been done into primary and secondary forms by taking into account the nature of their origin. Primary MPs are those which are produced at a tiny size and are commonly present in cosmetic products, drug materials and textiles [11, 12]. While UV-radiation and mechanical abrasion triggered breakage of larger plastic debris to produce fragmentation products are known as secondary MPs [13]. Secondary MPs are mainly found in industrial raw substances, domestic products, fishing nets, films, and other waste plastic remains [14]. MPs are not only an aesthetic issue, but also cause several other serious problems and impact various organisms due to their extended lifecycle, widespread presence all around the habitats, and minute size.

Furthermore, noxious chemicals which are released by the MPs, are utilized as plastic additives during the manufacturing and accumulate organic pollutants in them from the adjacent surroundings. Upon gaining entry into the food chains, these accumulated contaminants in MPs are then transferred to higher trophic levels from the environment [15]. Additionally, these MPs accumulated contaminants also cause acute toxicity in agronomic products and ultimately pose serious problems to human health [16].

MICROPLASTIC AS STRESSOR

Aquatic Environment

Different organisms experience different exposure to MPs such as organisms present at the bottom of the water surface encounter denser MPs like that of PET (polyethylene terephthalate) and PVC (polyvinyl chloride), while organisms present at surface waters come across those MPs which are less dense than seawater, for instance, PS (polystyrene), PP (polypropylene), and PE (polyethylene) [17]. The effect of MP fibers was studied in terms of the physiological and reproductive outcomes in pacific mole crabs *(Emerita analoga)* and mortality, fecundity, and embryonic developmental rates of crabs kept under control and microfiber-exposed conditions were compared. It was observed that the exposure to MPs increases with the increase in plastic usage and their addition to aquatic streams. The mortality rate was increased and embryonic development was also affected by an increase in the number of embryonic stages by MP microfibers [18].

MPs are also ingested, accumulated and transported into the hemolymph of Sydney rock oysters (*Saccostrea glomerata*) through microfold cells [19]. Microalgae, which are the primary producers of the aquatic system get impacted by MPs through a decline in their development and photosynthesis [20]. Alterations in the morphological parameters, a decline in the content of chlorophyll, and over-accumulation of ROS, were observed in microalgae by MPs [21]. The liver of fishes gets damaged*via* their exposure to MPs causing high lipid accumulation, inflammation, and also by a disturbance in lipid and energy metabolism [22]. Expression of inflammation-related and oxidative stress genes was increased, while a reduction in swimming competence and predacity along with the ability to locate the food *i.e.* population vigor; was also observed in young fishes [23]. In Zebra fishes, MPs exposure caused gut impairment and changes in the gut metabolome and microbiome [24]. Both direct and indirect toxicities can be imposed by MPs. Moreover, assessment of the interactive toxicity potential of MPs is also necessary as they can interact with other contaminants in the ecosystem [25].

Survival of lugworms (*Arenicola marina*) was found to be decreased due to the inhibition of their feeding activity and energy reserves by their exposure to MPs [26, 27]. Unusual swimming behavior in the goby (*Pomatoschistus microps*) was observed, whereas in Asian green mussel *Perna viridis*, aspects like the rate of respiration, production of byssus and clearance of food were decreased *via* their exposure to MPs [28, 29] which ultimately famish them to death in both the organisms. MPs ingestion in oysters markedly caused inhibition of the process of gamete formation and degradation in gamete quality, thus negatively hampering the overall frequency of fertilization, finally resulting in the reduced tally of oocytes and the rates of sperm production and alterations in their mantle or shell structures [30, 31]. Exposure of 50-nm amino-modified PS MPs to the embryos of the sea urchin (*Paracentrotus lividus*) caused thickening of the ectodermal membrane and exhibited irregular development at 6 h post-fertilization (hpf). Malformation in the larva, such as imperfect or absence of skeletal rods, decreased arm length and damaged ectoderm, was also observed at 48 hpf [32]. Ingestion of MPs induced aberrations in the ultrastructure of the intestinal epithelial cells of brine shrimp *Artemia parthenogenetica* and zebrafish *Danio rerio*, *via* causing a reduction in the number of microvilli, elevation in the mitochondria number, and separation of villi and of enterocytes, respectively [33, 34].

MPs are also known to alter gene expression at the molecular level. For instance, in *Sparus aurata* (gilthead seabream) and *Caenorhabditis elegans* (a nematode), the expression of stress response-associated genes was increased *via* their exposure to MPs [35]. Expression of oxidative stress genes was increased due to the over-accumulation of ROS, consequently resulting in the induction of the MAPK signaling pathway in juveniles of the Chinese mitten crab (*Eriocheir sinensis*) *via* MP exposure [36]. MPs exposure to the coral *Pocillopora damicornis* caused up- or down-regulation of expression of zymogen granules, transport of sterols, and the genes associated with JNK and EGF-ERK1/2 signaling pathways, suggesting that, physiological activities of this particular species were modulated by MPs *via* functioning as an environmental hormone [37]. Expressions of genes dealing with immunity, biotransformation, DNA repair, stress-induced responses, and lipid metabolism signaling pathways, were altered by MPs in the Mediterranean mussel *Mytilus galloprovincialis* and European bass fish *Dicentrarchus labrax* [38, 39].

MPs exposure to larvae of the *D. rerio* also caused alterations in gene expression [40].

Mortality and cytochrome P450 were induced in *D. labrax* by MPs exposure [41] while the growth and fecundity of freshwater crustacean, the amphipod *Hyalella Azteca* were impacted by PE MPs [42]. PE MPs exposure to marine mussel *Mytilus galloprovincialis* induced numerous noxious effects such as over-accumulation of ROS, disturbance in immune response, and genotoxicity [43]. Moreover, PE MPs caused an elevation in the energy consumption in polychaete *A. marina* [44]; whereas in echinoderm *Tripneustes gratilla*, larval growth and development were affected with no effect on its survival [45]. Nobre *et al.* [46] verified these results in larvae of green sea urchin *Lytechinus variegatus* (24 h;

200 ml particles L^{-1}). MP concentrations that have been practiced in these reported studies were higher as compared to those concentrations that are normally experienced in the aquatic system.

PS MPs exposure for 288 h to 0.05–6 μm; and 0.1–20 mg L^{-1} in rotifers resulted in reduced growth rate and fecundity (Jeong *et al.*, 2016). Exposure of 1 μm PS MPs at 50 mg L^{-1} concentration for 6 h to 24 h to scleractinian coral *Pocillopora damicornis*, reduced the activity of glutathione S-transferase (GST) and alkaline phosphatase along with triggering the functioning of antioxidant enzymes. Coral transcripts which were up-regulated were found to be mostly associated with the stress-induced responses and JNK signaling pathway, whereas the EGF-ERK1/2 signal pathway and transport of sterols were carried out by downregulated transcripts as revealed by RNA sequencing. Overall, it was concluded that coral stress response can be stimulated by acute MP exposure while JNK and ERK signaling pathways repress detoxification and the immune system [37]. Activities of catalase (CAT) and glutathione peroxidase (GPOX) were observed to be modulated as well as the content of neurotransmitter dopamine was also increased in circulating fluids of freshwater zebra mussels (*Dreissena polymorpha*) which was exposed to PS MPs (1 and 10 μm) for 6-days [47]. The injection of 700 nm PS MPs at a concentration of 5 mg mL^{-1} in embryos of zebrafish caused the stimulation of immune-associated gene products [48].

The activity of lactate dehydrogenase (LDH) and isocitrate dehydrogenase (IDH), which are energy-related enzymes, was changed while acetylcholinesterase (AChE) activity was suppressed as well as an increase in the lipid oxidation (LPO) was observed in the brain and muscle by exposure to MPs. Neurotoxic effects of MP have been reported in different species, although the particular mechanisms of MPs modulated neurophysiology of fish are not clarified [49, 50]. Sub-lethal effects were observed on embryo and larval development in *M. galloprovincialis* by their exposure to 3 μm PS MPs. But clearance rate and edible food intake were not impaired by MPs, despite the presence of MPs in the digestive tract over 192 h [51]. All of this implies that upon ingestion of MPs, life forms dwelling in aquatic ecosystems suffer in every aspect, in terms of their morphology, reproduction and other processes of vital importance.

Terrestrial System

Extensive studies have been performed in the aquatic system for MPs evaluation, whereas data in terrestrial systems which have a high content of plastics and their additives, is still limited. MPs pollution is very high in the terrestrial system due to the presence of anthropogenic activities and different other input ways [25], among which major activities are sewage-treated sludge [34], tire wear [52], packaging, use of plastic mulches, and atmospheric deposition [53].

Biodiversity and physical characteristics of soil are highly affected by the MPs after their accumulation and persistence in the environment [3]. Soil attributes like aggregation, microbial activity, water retention ability, and overall bulk density are influenced by the presence of MPs [54], and also change the soil's biogeochemistry, associated with plastic mulching [55]. MPs occurring in the terrestrial system can be ingested by terrestrial birds [56] or freshwater birds [57] and also cause a decline in the development of earthworms [58]. Among livestock, Panebianco *et al.* [59] found MPs in edible terrestrial snails and farm snails. Huerta Lwanga *et al.* [58] found an increase in the MPs concentration as transferred from soil to the chickens, indicating the trophic level transfer of MPs down the terrestrial food chain. Moreover, the consumption of such chickens having MPs accumulated in them can harmfully affect the health of human beings [60]. The MPs reach top consumers of the higher trophic level after their entry into the food chain and can even cross blood–brain barriers and trigger cognitive alterations or diseases according to their size [61]. MPs accumulation kinetics and distribution patterns in various tissues are determined by their particle size. Neurotoxicity in mice was induced by MPs *via* altering the lipid metabolism and energy-related processes and through inducing oxidative damage [62]. Health risks like cytotoxicity, undesirable immune response, hypersensitivity, and acute response like hematolysis, *etc.* are caused by the accumulation of MPs [63]. Intestinal cell damage, oxidative harm, and DNA impairment were induced by MPs in earthworms, even at low concentrations [64].

A similar level of MPs is accumulated in terrestrial or continental food webs as that of the marine system or even at higher levels. In China, the alimentary canal of 94% of deceased terrestrial birds contained MP with different foraging actions [65]. Agricultural activities are one of the major sources of release of MPs in the terrestrial system and moreover, MP has also been reported from the guts of freshwater continental birds [66, 67]. Ingestion of MPs in birds can be either unintentional or *via* trophic transfer in some cases, which was attributed to the smaller size of MPs as compared to the usual food of those birds [65]. Increased MP concentration of ~0.9, ~14, and ~129 particles g^{-1} was found in the soils, earthworm casts, and chicken feces which was the first quantitative evaluation of trophic level transfer of MP [68]. Intestinal lymphatic cells are the major tissues from which particles enter the body of humans, dogs, rabbits, and small rodents [69]. Though there is a long way to go in finding out more comprehensive details about the impact of MPs on the terrestrial ecosystem but the impairment of soil attributes and trophic level transfer are undeniable proof of their hazardous nature.

Microbes

Studies related to the influence of MPs on the transference and deposition of soil microbes are quite limited. In quartz sand, there was a negligible effect on bacteria *E. coli* by PS particles under low ionic strength circumstances, while stimulated bacterial transport was observed by plastic particles under high ionic strength conditions. It was concluded that at the nanoscale (20 nm), plastic-induced increased cell transport was chiefly due to the surface-level adsorption of plastic particles onto the cells and the repelling effect, whereas the competition for deposition of plastics on the sand at the microscale (2 μm) was considered mainly responsible for enhancing cell transport. However, the mechanism of movement of microbes under the influence of MPs in real soil systems requires further research [70].

A positive impact was exerted by the MPs on the microbial community in the soil rich in fertilizers [71]. It was suggested that dissemination and deposition of microbial diseases are mainly performed by the MPs which is attributed to the fact that the presence of pathogens like *Escherichia coli, Pseudomonas*, and human pathogens (*Bacillus cereus, Stenotrophomonas maltophilia*) in biofilms coated on the surface of MPs, is significant in the colonization of plastic-containing household items in aquatic systems [72]. Viršek *et al.* [73] found that the bacterial fish pathogen *Aeromonas salmonicida* is disseminated by MPs in the marine system, hence MPs are recognized as a potential vector. Algal characteristics like growth, colony size and morphology, photosynthesis, and gene expression, were altered by the presence of MPs, *via* passage of adsorbed contaminants from MPs, showing that MPs can accumulate in microalgae also [74]. Growth of marine microalgae *Skeletonema costatum* was curbed with the upper limit inhibition value of 39.7% *via* their exposure to micro-PVC (average diameter 1 micron) after 96 h exposure [75].

MPs' particles *i.e.* PFs (0.05-0.4%), polyacrylic (0.05-0.4%), and PS (polystyrene) (1 mg kg-1) in the soil exhibited adverse impacts on microbial communities [76, 77]. Bacterial growth in the soil was positively affected by their exposure to PP (polypropylene) [78]. Modifications in the soil structure were the major reason behind alterations in the activity of microorganisms in these studies. The growth of plants is directly impacted by the presence of microbial populations in the soil such as arbuscular mycorrhizal fungi (AMF), and diazotrophic bacteria. Plant and crop growth could be directly impacted *via* MPs-induced decreased microbial activity [79]. Moreover, the relative abundance and community profile of AMF were adversely affected owing to the presence of MPs in the soil [80]. PLA MPs were also observed to influence communications among microbial species occurring integrally in the soil [81], which in turn impact rates of microorganism-mediated processes such as absorption of minerals and fixation of nitrogen.

Soil enzymes for instance, urease, glucosidase, and phosphatase are also affected by the presence of MPs [82]. Activities of enzymes like acid phosphatase and urease in the soil were positively affected by MPs such as PE (polyethylene) and PVC (polyvinyl chloride) [83]. Congregation of unique bacterial species is a great ability of MPs as compared to the neighboring soil particles. There are structural changes in bacterial communities that dwell on MPs as compared to the circumambient MPs, plant refuse, and soil particles. MPs provide suitable sites for the colonization of various microbes as the surfaces of MPs like flakes and pits offer sites for active hydrolysis which facilitate their colonization, hence, MPs offer different ecosystem habitats for soil bacteria [84].

Chloroflexi, Acidobacteria, Bacteroidetes, and Gemmatimonadetes were the bacterial community composition which were found to be actively colonizing MPs in farmland soils [85]. Among bacterial communities, these bacteria are called as "Keystone Species" and MPs are recognized as "special microbial accumulators" [85]. Specific microbial communities comprising disease-producing and plastic-degrading bacteria get unique niches from MPs and these microbial communities can degrade MPs and can also change the biological activities of the soil ecosystem [86]. MPs surfaces were also observed to be a site provider to species like *Arthrobacter*, *Streptomyces*, *Nocardia*, *Aeromicrobium*, *Janibacter*, and *Mycobacterium* in the soil [87]. It can be concluded that MPs not only act as a carrier of microbial pathogens and alter the composition of microbial flora in the soil but also get decomposed by some special bacterial species.

Plants

Can plants uptake and accumulate MPs, and how are plant growth and food quality are impacted by MPs, are the two most common questions that come in people's minds. Right now, information on this subject is scarce, most probably due to the fact that it is tough to recognize MPs in plant tissues. It is likely that small-sized MPs are able to cross cell wall and plasma membrane barriers. Fluorescent microbeads can be utilized to probe the uptake of microplastics by plants. Tobacco cells could be invaded by nano-scale (< 100 nm) fluorescent PS beads through the process of endocytosis as demonstrated by the cell culture-based study [88]. Micro-scale (0.2 μm) fluorescent PS beads could be accumulated from the soil by edible plants as studied in a wholeplant culture [89], suggesting the MPs mediated health risks to humans through the food chain. Growth of wheat was disturbed by the addition of 1% biodegradable plastic particles and PE plastic particles in the soil, in which, former MP showed a more negative impact on wheat as compared to the latter. Moreover, biodegradable plastic particles exhibited a negative impact on plant biomass. Interestingly, MPs-induced damages were alleviated by the presence of earthworms in the plastic [90].

The inclusion of MPs, either singly or in combination, affects different plant species in a community to a varying degree and they also possess the ability to influence their composition and diversity. Soil structure characteristics like soil aggregation, are associated with the plant community properties [91, 92]. Hence, various types of MPs induced significant effects on soil properties that have the ability to alter the composition of various plant communities. Soil water evaporation was increased by plastic films and resulted in more pronounced water deficit conditions and consequently improved the development of drought-tolerant plant species in a community [93]. Plant community composition, productivity, and diversity are strongly influenced by the soil microbial community [79, 94, 95]. MP addition-mediated alterations in soil microbial composition or root-colonizing symbionts may further impact the composition of the plant community. For example, if microbial diversity of the soil or the abundance of root-colonizing symbionts is decreased by the addition of MP, then there is a reduction in the plant diversity, as plant diversity is positively correlated with the microbial diversity of the soil or root-colonizing symbionts [94, 95]. Such impacts on plant diversity are more likely to arise in those areas where MP pressure is greater and this problem is of high concern near agronomic fields or towns.

Physical and chemical variables of the soil change in the presence of MPs, which then results in an altered root system and vegetative period to reduce the functioning of plants [96, 97]. Rhizosphere microbes such as *Azotobacter* and pathogenic and mycorrhizal fungi influence the growth of plants [98]. Plant performance is indirectly affected by MPs as MPs' existence in the soil changes the soil properties, which in turn alters the relevant microbial activities and the structure of microbial communities.

MP microfibers contamination in soil caused a reduction in soil bulk density due to which soil aeration increased which may benefit fromhigher root penetration into the rhizosphere see Fig. (**1**) [99].

In addition to this, juvenile roots can be entangled by microfibers and seedlings' growth could be hindered. Moreover, the dryness of the soil surface can be increased by elevating the aeration of the soil which is also responsible for negatively impacting the growth and development of the seedlings. Water is deeply percolated to deeper soil layers *via* water channels created by MP thin films and also triggers the creation of desiccation cracks in the soil surface [93].

Reduction or improvement in the cohesion between –aggregate-forming soil particles is one of the MPs-induced altered physical properties of soil. In this regard, the process of formation of aggregates is the most important contribution of microfibers. It was reported from a study conducted in soil that 72% of MPs take part in aggregate formation [100]. Lozano and Rillig [101] observed that soil polluted with microfiber caused alterations in the structural composition of the plant community and exhibited that plant species dominance could be affected by MPs because, in microfiber-rich soil, a European invasive species *Calamagrostis*, and an allelopathic species *Hieracium* became dominant. But healthy plant community facilitated by plant species such as *Holcus* (velvet grass) was observed to have a reduction in biomass. Secondary effects such as drought induced by microfibers could be responsible for these effects at the plant community level. Exposure of LDPE MPs to winter ryegrass *Lolium perenne* caused a decrease in their root growth [102]. However, they were unable to find the reason behind this change-whether this is on account of alterations in physical attributes of the soil or simply a result of some other pernicious chemical characteristics of MPs?

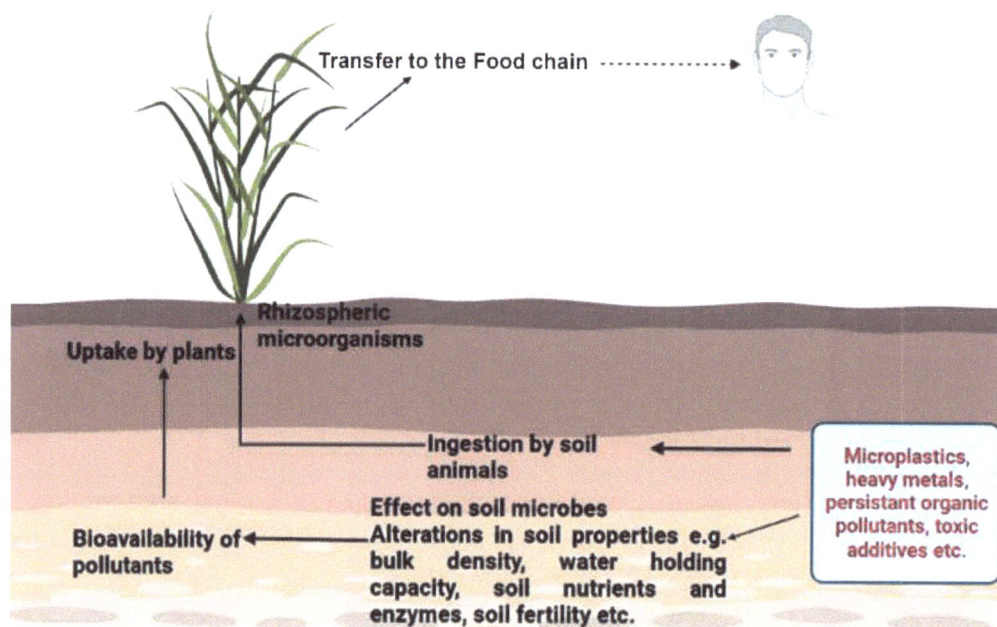

Fig. (1). Diagrammatic representation of effect of MPs on soil properties and their transfer to the food chain.

Exposure of plastic particles at three different sizes *i.e.* at 50, 500, and 4800 nm in a concentration range of 103-107 particles ml^{-1} resulted in a 78% decline in the seed sprouting process of *Lepidium sativum* (garden cress) after exposing them for

8h [103]. This reduction in the germination of seeds was attributed to the MPs induced physical clogging of the seed pores which got confirmed with the usage of confocal microscopy and fluorescent dyes. MPs were also accumulated in root hairs and after degradation, these MPs transformed into plastics having even smaller dimensions called nano plastics. It was shown by some recent studies that MPs can be accumulated by terrestrial plants. MPs of 100 nm size were found to be accumulated in *Vicia faba* (Fava bean) roots as evaluated with laser confocal scanning microscopy due to which cell wall pores or cell junctions which are helpful for the transport of water and other nutritive materials, were blocked [104]. Accumulation of MPs in plants significantly improved the activities of antioxidant enzymes and reduced plant growth. In wheat (*Triticum aestivum*), fluorescently labeled MPs were observed to be localized in the whole plant tissues. According to a confocal laser scanning microscope study, PS MP microbeads were observed in the parts of the plant encompassing all from the root, and shoot to leaves [105].

Li *et al.* [106] studied the absorption, distribution, and dissemination of fluorescent polystyrene microbeads (0.2 mm) in *Lactuca sativa* (lettuce), a member of the Asteraceae family and observed that these polystyrene microbeads were entangled in the extracellular region of the root cap, which was visible to the naked eye as a dark green tip. Microbeads were localized in the cell wall and intercellular spaces of the root and vascular system through apoplastic transport and transpiration stream respectively, as shown by confocal microscopy. These microbeads combined with each other in both locations to make chains and clusters like organizations. Meanwhile, microbeads present inside the leaf were observed to be homogeneously distributed. A reduction of 83.3% was observed in the root growth of *Vigna radiata* (Mung bean) when exposed to nano-plastic polluted soil at a concentration of 100 mg kg^{-1} [107]. Moreover, nano plastic particles got accumulated inside the foliar parts of *V. radiata*. The growth rate of the African giant snail *Achatina fulica* when grown on this plant got badly affected by this contaminated mung bean plant. Therefore, it was suggested that whether MPs are absorbed by the plant or simply adhere to their surface, they are prone to be consumed by the primary consumers *i.e.* herbivores with their plant food. There is an immense need to study the *in vivo* impacts of MPs on various edible crop plants.

MICROPLASTIC AND OTHER CONTAMINANTS

MPs particularly NPs, on account of their small size markedly impact the nature of organic pollutants in the ecosystem *via* a dynamic balance of sorption and desorption [108, 109]. Polycyclic Aromatic Hydrocarbons (PAHs) are the most prevailing contaminants existing in the ecosystem together with MPs among all the organic contaminants, having contents in seawater, beaches, and sediments as high as 119 $\mu g\ g^{-1}$, 44.8 $\mu g\ g^{-1}$, and 5615 mg kg^{-1} MPs, respectively [110 - 112]. Different organic contaminants adsorbed on MPs could get easily desorbed within 24 h when exposed to clean seawater [113], suggesting the high hazard of these contaminants in the marine environment. Hence, it is significant to decipher the mechanism of interaction between MPs and organic contaminants and the ecological hazards of the sorption and desorption of these organic contaminants on or from MPs. The most commonly investigated MPs type was Polyethylene (PE) followed by Polystyrene (PS) among all the studies performed on the sorption and desorption of organic contaminants on or from MPs/NPs, to which size range of 150-280 μm being the one mostly offered. However, the availability of only a few isolation and estimation techniques for NPs has resulted in limited research works centered around the sorption of pollutants onto the surface of nano-MPs.

Sorption and accumulation of both organic and inorganic pollutants by MPs are possibly harmful as the accumulated pollutants can be dispersed into the ecosystem and can enter into the alimentary canal of various organisms [114, 115]. Pollutants like dioxin-like chemicals, heavy metals (Ni, As, Pb, Cd, Cr, *etc.*), polycyclic aromatic hydrocarbons (PAHs), polychlorinated biphenyls (PCBs), polybrominated diphenyl ethers (PBDEs), various pharmaceuticals products, *etc.* were reported to get adsorbed onto the surface of MPs [116].

Deanin [117] reported that plastic polymers are intermixed with monomeric particles or additives, not used in their pure forms, to increase the characteristics of the particles. Humans and other animals have been reported to be influenced by the presence of additives like phthalates in plastic and such additives were found in human urine and blood samples [118]. Nonylphenol and brominated flame retardants are the types of plastic additives that have been reported as carcinogens and inducing endocrine-related disorders. Organisms are susceptible to ingesting MP particles due to a decline in their size and also possess toxic moieties which are practiced as additives or accrued on MPs owing to their high adsorption potential for MPs, and easily enter inside the body of various organisms. Even starvation is caused by the MP *via* obstructing the nutritious adjuncts, and toxicity is also caused by the different toxic chemicals adhered to the surface of MPs. Upon their ingestion, MPs can bring about both physical and chemical changes in the neighboring organisms, therefore MPs act as multiple stressors.

Along the Persian Gulf, the interaction of MP with metals was studied and it was reported that metals with varied concentrations gathered on the surface of MPs. High risk was posed to the aquatic organisms by the MP in association with adsorbed metal content, which may further affect the human body indirectly by entering into the food chain. The process of adsorption may further enhance the aging of plastic *via* the formation of more reactive and heterogeneous surfaces [119]. Each kind of composite has its own specific properties, and due to the formation of new adsorption bonds, there is an alteration in the adsorption mechanism. An additional pathway is offered by these adsorption bonds for the adsorption of chemical pollutants onto the MP surface. MP particles were known to accumulate chromium on their surface under experimental circumstances and chromium was effectively adsorbed on the polyethylene microbeads in the seawater. Surface area and metal reactivity toward the MP particle are the major factors for the adsorption rate. Heavy metal enters aquatic organisms *via* the ingestion of metal-adsorbed MP, which is an additional pathway for their entry [120]. Metals with different concentrations were reported to accumulate in MP pellets, gathered up from 19 seaside areas along the coast of Sao Paulo state (Brazil). MP pollution is mainly caused by anthropogenic activities and is also harmful due to its potential to accumulate noxious metals [121].

Good linear correlations were shown by the partition coefficients of hydrophobic organic compounds (HOCs) such as the Polycyclic aromatic hydrocarbons (PAHs) and Hexachloro cyclohexanes (HCHs) with R2 values of 0.92, 0.94, and 0.84 on PE, PP, and PS MPs, respectively [122]. Hydrophobic forces with MPs are of high significance as indicated by the noteworthy association between the hydrophobicity of HOCs and partition coefficients [123]. In comparison to other mechanisms like electrostatic interaction and H-bonding, hydrophobic partitioning is recognized as a major mechanism for the adsorption of nonpolar organic moieties onto MPs [123, 124]. Nonpolar organic moieties display higher sorption to MP particles due to their higher hydrophobicity as compared to polar compounds [125].

In the case of polar organic moieties, hydrophobic nature is not exclusively the only regulating facet of sorption, with LogKow less than 2 and a varying range of acid dissociation constants (pKa). pH is another important factor that significantly alters the sorption of existing species of different organic compounds of polar nature on MPs. More interactions such as electrostatic interactions are involved between MPs and organic compounds. Further, the impact of existing species is also linked with the charge present on the surface of MPs [126, 127]. The sorption of charge-neutral species is stronger as compared to the charged species [125]. For instance, as a result of similar charges present on the surface of MP particles (PE, PP, and PS) and tetracycline, there was a significant inhibition of the sorption of charged tetracycline of either anionic or cationic nature on MPs due to the existence of electrostatic interactions [128]. Zhang *et al.* [129] studied a similar process in which the sorption of charged oxytetracycline was evaluated onto MPs. Guo *et al.* [126] found under acidic conditions that higher sorption affinity was shown by the positively charged tylosin to PS and PVC MPs in comparison to the neutral species, which was attributed to the electrostatic interactions formed on account of the opposite charges present on MPs and tylosin.

The sorption of organic moieties to MPs is also impacted by the presence of organic matter. Fluorescence measurement has confirmed that there is an insignificant association between MPs and dissolved organic matter (DOM; *e.g.*, humic acid) [130], which causes competitive accumulation of HOCs between MPs and DOM. In the presence of DOM between a concentration range of 150–1000 mg L^{-1}, the separation of phenanthrene and tonalide was observed to be altered between MPs and water [130]. DOM at a low concentration of 20 mg L^{-1} is observed to affect and inhibit the sorption of polar organic composites [131, 132] and also plays an opposing role in the accumulation of organic moieties in aged MPs. The occurrence of humic acid was observed to enhance the sorption of oxytetracycline on weathered PS MPs due to the formation of the complex between humic acid and weathered PS. This is attributable to the fact that humic acid functions as a connecting link between the exterior of aged MPs and oxytetracycline [129]. These MP particles can also be associated with the aromatic organization of humic acid through p-p conjugation when virgin MP size is decreased to the nanometer scale [133].

Pollutants like persistent organic pollutants (POPs) and organochlorine pesticides are also reported to accumulate on MPs [134], which involve hexachlorocyclohexane isomers (HCHs), polycyclic aromatic hydrocarbons (PAHs), polychlorinated biphenyls (PCBs), and the insecticide dichlorodiphenyltrichloroethane (DDT) [135]. MPs higher than 18 000 ng g^{-1} have been found for PCBs while for PAHs, these MPs are in the concentration range of 164 900 ng g^{-1} [136]. In addition to this, other toxic POPs like DDTs and HCHs [135] need a much lower concentration of MPs. Coupling of POPs due to their adsorption potential to MPs causes high dispersion and reduces their effect on the environment due to the resistance potential of POPs to degradation cascade of biochemical and photolytic nature [137]. Hence, besides having well studied the direct effect of MPs on aquatic biota, it was shown by various studies that the accumulation of pollutants on MPs could possibly have a higher effect on the ecosystem and the food chain.

MICROPLASTIC AS MULTIPLE STRESSORS

Evaluation of the combined noxious impacts of MPs and other pollutants in association with probable particle ingestion mediated physical damage have similarities to methods practiced for analyzing the hazards of chemical combinations *i.e.*, different mode-of-action exhibited by multiple stressors. During the process of addressing damages from chemical combinations, the major purpose is to assess the collective impacts of more than one stressor and to evaluate whether associations among the involved stressors result in mixture-specific impacts which derive from additives *i.e.* synergistic or antagonistic association [138]. A number of stressors impacting various ecosystem organisms have been increased in the environment through human-triggered changes.

In aquatic ecosystems, MPs and temperature stress have become the utmost evaluated stressors due to the alteration in climatic conditions and plastic pollution. In this context, Weber *et al.* [139] investigated the combined effect of PS MPs (6.4, 160, 4000, 100,000 p mL^{-1}) at either 14, 23, or 27 °C on zebra mussel *Dreissena polymorpha* to check the death rate, mussel functioning and clearance rate, energy storage, oxidative damage, and the immunological response. Activity, energy storage, oxidative stress and immune response were found to be affected by the rise of water temperature and showed a significant impact on *D. polymorpha*. However, only anti-oxidative potential was altered by the MPs with no communication impacts between MPs and temperature exposure. The study also revealed that a minor impact was shown by MPs on freshwater mussels as compared to temperature stress and the adaptation mechanism of dreissenids to suspended solids could be the reason for this limited MPs-induced toxicity in *D. polymorpha* [139].

The filtration ability and survival of *Daphnia magna* were studied through their exposure to water temperature, ammonium and PS MPs. It was found that its filtration potential and survival are greatly affected in a negative manner *via* the synergistic action of these three stressors *i.e.* water temperature, ammonium and MPs. Reduction in the range of 4.8% to 54.5% in TD50 (time for 50% mortality) and a decline of 13.1% to 91.7% in the t50% time (the time required for 50% decrease in the filtration ability) were observed in *D. magna* by the combined effects of three stressors vis-à-vis subjection to only two stressors [140]. Combination of Ni and MPs exhibited toxic effect on *D. magna*, suggesting the synergistic association between Ni and MPs [141]. Effect of ambient (~27 °C) and elevated (~30 °C) temperature was studied on two coral species, rice coral (*Montipora capitata*) and lace coral (*Pocillopora damicornis*), and then fed MPs, brine shrimp (*Artemia nauplii*), or both. It was found that feeding was significantly decreased on *Artemia* by both species but there was no significant reduction in the ingestion of MPs after exposure to temperature stress [142].

Kratina *et al.* [143] investigated the individual and synergistic effects of MPs and temperature on the metabolic and ingestion rates of a model freshwater detritus eater. MPs-induced effects on metabolic rate were altered by experimental warming, and at the lowest temperature, this metabolic rate was observed to be elevated with MP concentration while reduced at the higher temperatures. It was suggested from the study that short-term exposure to MPs altered the metabolism of important freshwater detritivores, with high temperature triggered inhibition of metabolic rates [143]. Different species show different sensitivity to various types of MPs and could be considerably impacted by temperature even also at greater exposure concentrations [144]. Several biological effects such as impairment in feeding and the production of mucus to alterations in gene expression were found in corals by ingesting polypropylene when exposed to MPs. The coral microbiome was altered by MPs in direct as well as indirect manner by causing tissue bruises which permit the development of opportunistic bacteria. These MPs-induced multifold impacts indicate that at present, MPs at the concentrations existing in some marine regions and oceans can eventually result in coral death [145].

Individual and combined impacts of PS MPs *i.e.* of 1-µm and 10-µm size and roxithromycin (ROX) were studied on a planktonic crustacean *Daphnia magna*. The activity of CAT and GST enzymes and MDA levels were found to be activated when exposed to PS (0.1 mg L^{-1}) or ROX (0.01 mg L^{-1}) alone for 48 hours. 1-µm PS exposure caused the decreased activity of GPx and MDA in *D. magna* if compared with its subjection to ROX alone, whereas the activity of GST and MDA was decreased by the co-exposure to 10-µm PS. Thestrongest biological responses were caused in *D. magna* by co-exposure to 1-µm PS and ROX. The activity of SOD was highly stimulated in the PS and ROX co-exposed in daphnids in comparison to the individual treatment of ROX. These enhanced results may be due to the adsorption of ROS on PS, resulting in greater stockpiling of ROX in *D. magna* than ROX alone [146]. Combined application of venlafaxine and MPs caused a threefold increase in the SOD activity as induced by venlafaxine alone [147]. Similarly, combined exposure to MPs and cephalexin resulted in post-exposure predatory performance in common goby fish *Pomatoschistus microps* as compared to their individual effects [148]. Thus, the individual effects of MPs vary greatly when compared to their synergistic role in combination with other stressors such as temperature, heavy metals and antibiotics.

MICROPLASTIC TO NANO PLASTIC

Nano plastics exhibit different properties from MPs as nano plastics are much smaller in size. 45 minutes exposure of PS nano plastics (100 nm) on Blue mussel (*Mytilus edulis*) and Eastern oyster (*Crassostrea virginica*) for their ingestion and egestion was observed and it was found that nano plastics can be passaged to the digestive gland and persist in the body for more time as compared to the MPs [149]. Extraction and accumulation of exopolymeric substances (EPS) by three species of phytoplankton were observed by using 10 and 100 ppb of PS-engineered nanoparticles of non-fluorescent and fluorescent nature (both 23 nm). It was concluded that hydrophobic interactions of engineered nanoparticles with EPS polymers were the major reason behind engineered nanoparticles' induced effects on EPS assembly kinetics [150]. The generation of pseudo feces and the weight of feces compared to that of pseudofeces were analyzed through PS nano plastics' (30 nm) exposure to *M. edulis*. Nano plastics were significantly removed from the water by *M. edulis* and exhibited decreased filtering activities in the proximity of algae (food) [151].

Sea urchins (*Paracentrotus lividus*) exhibited embryotoxicity and anomalies in gene expression by their exposure to PS nano plastics with functional groups like carboxyl (40 nm) and amine (50 nm) [152]. Amino-modified polystyrene (PS-NH$_2$) nano plastics (50 nm) were used *in vitro* to evaluate various aspects such as the stability of lysosomal membranes, the release of lysosomal enzyme, phagocytosis, production of extracellular oxyradicals, *etc.* in *Mytilus galloprovincialis* Lam and it was found that PS-NH$_2$ affected different parameters of the immune system and brought about pre-apoptotic developments [153, 154]. Mortality rates and sub-lethal effects on a species of brine shrimp *Artemia franciscana* were analyzed through their exposure to carboxylated and amino-modified PS nano plastics (40–50 nm) [155]. These nano plastics adversely affected the activities such as feeding, movements, and multiple molting in *A. franciscana* and also affected the organism, population and ecosystem.

Mortality, food intake, sex ratio, and reproductive potential of a copepods*Tigriopus japonicus*were studied by using both micro- and nano plastics (50 nm, 500 nm, and 6 µm) of plain and fluorescent nature. Fertilization was physically inhibited by all PS beads except 50-nm PS beads which had no effect on the fertility of *T. japonicus* [156]. Larvae of pacific oysters (*Crassostrea gigas*) were evaluated on the basis of the size of plastics to check the consumption of micro- and nano-plastics, and MPs beads were also used to analyze the ingestion and growth. The age of the larva, size, and surface properties of plastics are the major factors for ingestion during a one-day long exposure. Larvae easily digested and retained laminated plastics in their bodies while feeding and growth of *C. gigas* were not significantly affected by the plastics [157].

Ingestion, egestion, generation of Reactive Oxygen Species, and activity of antioxidative enzymes of a rotifer *Brachionus koreanus* were evaluated by their exposure to fluorescently labeled PS micro beads at different sizes *i.e.* 50 and 500 nm and 6 µm) and also the reproductive potential and lifespan of *Brachionus koreanus* were observed. Factors such as growth rate, lifespan, fertility, reproduction time, and body dimensions of *B. koreanus* were found to be decreased by their exposure to plastic beads and also elevated oxidative stress and antioxidant enzymes. It was suggested that bigger plastics are less toxic as compared to smaller plastics and size-dependent toxicities are shown by the micro- and nano plastics [158]. Virgin fluorescent latex particles (50 and 500 nm) and carboxyl-group modified fluorescent latex particles (50 and 500 nm) were used to check the uptake, mortality, and excretion in embryos and yolk-sac larvae of Japanese rice fish (*Oryzias latipes*). It was concluded from the study that *O. latipes* ingested and excreted smaller nano plastics more easily and slowly, as compared to the bigger MPs [159].

Table 1. Microplastics: Sources and major effects on the biotic component of the Ecosystem.

Sources	Effect on Animals	Effect on Plants	Effect on Microbes
Wastewater treatment plants, landfills, fertilizers, irrigation, agricultural fields, textiles, industrial effluent, and domestic runoff coastal and tourism activities, commercial fishing, marine vessels, aquaculture, and oil rigs [162 - 169]. Subsequently, microplastic gets piled up in the deep sea, pristine polar regions, ice sheet, living beings, and contaminants such as polycyclic aromatic hydrocarbons (PAH), heavy metals, and antibiotics [170 - 172].	Adverse effect on fecundity and embryonic developmental rates [18], disturbance in lipid and energy metabolism [22], alterations of gene expression at various levels and stimulation of stress responsive genes [35], trophic level transfer, triggering of cognitive malformations due to crossing of blood-brain barriers [61].	Reduction in plant growth and biomass, low rates of seed germination [102], the entanglement of juvenile roots, negatively affecting the growth of seedlings, alters soil attributes and rhizospheric microbial flora causing drastic change in composition and diversity of plant communities, makes water availability scarce for the plants [93].	Act as a surface for microbial colonization but cast a negative impact on the relative abundance and community profile of microbial species in the soil such as Arbuscular Mycorrhizal Fungi (AMF) [80], disturb the rates of micro-organism mediated processes such as mineral absorption and nitrogen fixation [81], disseminate microbial pathogens.

PS micro- and nano plastics were used to assess their toxicities in zebrafish (*Danio rerio*). MPs of 20 µm size got piled up only in the gills and gut of *D. rerio* whereas the gills, liver, and gut were the different body parts in which 5 µm MPs accumulated after one-week long exposure. Inflammation and accumulation of lipids in the liver of *D. rerio* were observed by their exposure to both 5 µm and 70 nm PS-MPs as indicated by a histopathological study. Oxidative stress was also induced by PS-MPs by increasing the functioning of enzymes as superoxide dismutase (SOD) and catalase (CAT) [160]. Virgin PS beads of three different sizes *i.e.* 50 and 500 nm and 6 µm were used to evaluate their effect on photosynthetic ability and the growth attributes of the alga *Dunaliella tertiolecta* and it was found that adverse impacts were increased with the reduction in the particle size [161]. This discussion clearly indicates that greater toxic impacts were caused by smaller micro- or nano plastics on organisms which is attributed to their relatively huge surface area and other properties [158, 161].

Various sources of microplastics and their impact on animals, plants and microbes have been summarized in Table **1**.

CONCLUSION

MPs are small-sized particulates that are considered atmospheric contaminants and are known to pose serious hazards to the ecosystem. MPs-induced ecosystem damages will get more acute if plastic ingestion and its worldwide manufacturing are uninterruptedly increased. Organic contaminants induced toxicity to aquatic organisms is altered through MPs *via* different pathways, and "carrier", "releaser", "dilution" or "cleansing" effects exerted by these particles determine the elevation or decline in their bioaccumulation. Soil contaminated with MP is found almost everywhere which alters soil quality, soil organisms and various other properties of soil. Moreover, MPs accumulated noxious pollutants can be transferred with the migration of MPs, however, MPs do not have any key role in the transport of HOCs to soil-dwelling organisms such as earthworms.

Soil fertility is significantly regulated by soil biota but there is limited information regarding the interaction between soil-dwelling organisms and MPs. Toxic contaminants are released into the environment by MPs which pose noxious impacts on plants and soil biota. MPs-mediated alterations in plant growth and development exhibit species specificity. Primary production and terrestrial ecosystem functioning are adversely affected by MPs-induced negative effects on plant performance. Moreover, human health is also impacted by the sorption and transference of noxious pollutants accumulated with MPs. It is a prerequisite to predict and assess the comprehensive risk of MPs induced impact on various organisms therefore, future research should focus on the following aspects (i) for the accurate evaluation of sorption coefficients, there is a need for regression descriptors for MPs detected in the ecosystem (ii) the sorption and desorption studies of additives on MPs and of field-collected MPs. The elevation in the production of plastic materials, high sturdiness of plastic substances, destruction of prevailing plastic contamination and upsurge in the occurrence of MPs and nano plastics will lead to high plastic pollution which is expected to persist across the globe in the future as well.

ACKNOWLEDGEMENTS

Authors are thankful to Lovely Professional University, Punjab.

REFERENCES

[1] Rillig MC. Microplastic in terrestrial ecosystems and the soil? Environ Sci Technol 2012; 46(12): 6453-4.
[http://dx.doi.org/10.1021/es302011r] [PMID: 22676039]

[2] Lambert S, Sinclair C, Boxall A. Occurrence, degradation, and effect of polymer-based materials in the environment. Rev Environ Contam Toxicol 2014; 227: 1-53.
[PMID: 24158578]

[3] Wagner M, Scherer C. Alvarez-Mu~noz, D.; Brennholt, N.; Bourrain, X.; Buchinger, S.; Fries, E.; Grosbois, C.; Klasmeier, J.; Marti, T.; Rodriguez-Mozaz, S.; Urbatzka, R.; Vethaak, A.; Winther-Nielsen, M.; Reifferscheid, G.; Microplastics in freshwater ecosystems: what we know and what we need to know. Environ Sci Eur 2014; 26(1): 1-9.

[4] Barnes DKA, Galgani F, Thompson RC, Barlaz M. Accumulation and fragmentation of plastic debris in global environments. Philos Trans R Soc Lond B Biol Sci 2009; 364(1526): 1985-98.
[http://dx.doi.org/10.1098/rstb.2008.0205] [PMID: 19528051]

[5] Bläsing M, Amelung W. Plastics in soil: Analytical methods and possible sources. Sci Total Environ 2018; 612: 422-35.
[http://dx.doi.org/10.1016/j.scitotenv.2017.08.086] [PMID: 28863373]

[6] Horton AA, Walton A, Spurgeon DJ, Lahive E, Svendsen C. Microplastics in freshwater and terrestrial environments: Evaluating the current understanding to identify the knowledge gaps and future research priorities. Sci Total Environ 2017; 586: 127-41.
[http://dx.doi.org/10.1016/j.scitotenv.2017.01.190] [PMID: 28169032]

[7] Wilcox C, Puckridge M, Schuyler QA, Townsend K, Hardesty BD. A quantitative analysis linking sea turtle mortality and plastic debris ingestion. Sci Rep 2018; 8(1): 12536.
[http://dx.doi.org/10.1038/s41598-018-30038-z] [PMID: 30213956]

[8] Moser ML, Lee DS. A fourteen-year survey of plastic ingestion by western North Atlantic seabirds. Colon Waterbirds 1992; 15(1): 83-94.
[http://dx.doi.org/10.2307/1521357]

[9] Buxton RT, Currey CA, Lyver POB, Jones CJ. Incidence of plastic fragments among burrow-nesting seabird colonies on offshore islands in northern New Zealand. Mar Pollut Bull 2013; 74(1): 420-4.
[http://dx.doi.org/10.1016/j.marpolbul.2013.07.011] [PMID: 23899612]

[10] Hartmann NB, Hüffer T, Thompson RC, *et al.* Are we speaking the same language? Recommendations for a definition and categorization framework for plastic debris. Environ Sci Technol 2019; 53(3): 1039-47.
[http://dx.doi.org/10.1021/acs.est.8b05297] [PMID: 30608663]

[11] Browne MA. Sources and pathways of microplastics to habitats.Marine Anthropogenic Litter. Cham: Springer 2015; pp. 229-44.
[http://dx.doi.org/10.1007/978-3-319-16510-3_9]

[12] Cole M, Lindeque P, Halsband C, Galloway TS. Microplastics as contaminants in the marine environment: A review. Mar Pollut Bull 2011; 62(12): 2588-97.
[http://dx.doi.org/10.1016/j.marpolbul.2011.09.025] [PMID: 22001295]

[13] Rezania S, Park J, Md Din MF, *et al.* Microplastics pollution in different aquatic environments and biota: A review of recent studies. Mar Pollut Bull 2018; 133: 191-208.
[http://dx.doi.org/10.1016/j.marpolbul.2018.05.022] [PMID: 30041307]

[14]   Eerkes-Medrano D, Thompson RC, Aldridge DC. Microplastics in freshwater systems: A review of the emerging threats, identification of knowledge gaps and prioritisation of research needs. Water Res 2015; 75: 63-82.
[http://dx.doi.org/10.1016/j.watres.2015.02.012] [PMID: 25746963]

[15]   Ward JE, Kach DJ. Marine aggregates facilitate ingestion of nanoparticles by suspension-feeding bivalves. Mar Environ Res 2009; 68(3): 137-42.
[http://dx.doi.org/10.1016/j.marenvres.2009.05.002] [PMID: 19525006]

[16]   Huang Y, Liu Q, Jia W, Yan C, Wang J. Agricultural plasticmulching as a source of microplastics in the terrestrial environment. Environ Pollut 2020; 260: 260.
[http://dx.doi.org/10.1016/j.envpol.2020.114096]

[17]   Cole M, Lindeque P, Fileman E, *et al.* Microplastic ingestion by zooplankton. Environ Sci Technol 2013; 47(12): 6646-55.
[http://dx.doi.org/10.1021/es400663f] [PMID: 23692270]

[18]   Horn DA, Granek EF, Steele CL. Effects of environmentally relevant concentrations of microplastic fibers on Pacific mole crab (Emerita analoga) mortality and reproduction. Limnol Oceanogr Lett 2020; 5(1): 74-83.
[http://dx.doi.org/10.1002/lol2.10137]

[19]   Scanes E, Wood H, Ross P. Microplastics detected in haemolymph of the Sydney rock oyster *Saccostrea glomerata*. Mar Pollut Bull 2019; 149: 110537.
[http://dx.doi.org/10.1016/j.marpolbul.2019.110537] [PMID: 31466014]

[20]   Sjollema SB, Redondo-Hasselerharm P, Leslie HA, Kraak MHS, Vethaak AD. Do plastic particles affect microalgal photosynthesis and growth? Aquat Toxicol 2016; 170: 259-61.
[http://dx.doi.org/10.1016/j.aquatox.2015.12.002] [PMID: 26675372]

[21]   Prata JC, da Costa JP, Lopes I, Duarte AC, Rocha-Santos T. Effects of microplastics on microalgae populations: A critical review. Sci Total Environ 2019; 665: 400-5.
[http://dx.doi.org/10.1016/j.scitotenv.2019.02.132] [PMID: 30772570]

[22]   Lu Y, Zhang Y, Deng Y, *et al.* Uptake and accumulation of polystyrene microplastics in zebrafish (*Danio rerio*) and toxic effects in liver. Environ Sci Technol 2016; 50(7): 4054-60.
[http://dx.doi.org/10.1021/acs.est.6b00183] [PMID: 26950772]

[23]   Qiang L, Cheng J. Exposure to microplastics decreases swimming competence in larval zebrafish (*Danio rerio*). Ecotoxicol Environ Saf 2019; 176: 226-33.
[http://dx.doi.org/10.1016/j.ecoenv.2019.03.088] [PMID: 30939402]

[24]   Qiao R, Sheng C, Lu Y, Zhang Y, Ren H, Lemos B. Microplastics induce intestinal inflammation, oxidative stress, and disorders of metabolome and microbiome in zebrafish. Sci Total Environ 2019; 662: 246-53.
[http://dx.doi.org/10.1016/j.scitotenv.2019.01.245] [PMID: 30690359]

[25]   Horton AA, Walton A, Spurgeon DJ, Lahive E, Svendsen C. Microplastics in freshwater and terrestrial environments: Evaluating the current understanding to identify the knowledge gaps and future research priorities. Sci Total Environ 2017; 586: 127-41.
[http://dx.doi.org/10.1016/j.scitotenv.2017.01.190] [PMID: 28169032]

[26]   Besseling E, Wegner A, Foekema EM, van den Heuvel-Greve MJ, Koelmans AA. Effects of microplastic on fitness and PCB bioaccumulation by the lugworm *Arenicola marina* (L.). Environ Sci Technol 2013; 47(1): 593-600.
[http://dx.doi.org/10.1021/es302763x] [PMID: 23181424]

[27]   Wright SL, Rowe D, Thompson RC, Galloway TS. Microplastic ingestion decreases energy reserves in marine worms. Curr Biol 2013; 23(23): R1031-3.
[http://dx.doi.org/10.1016/j.cub.2013.10.068] [PMID: 24309274]

[28]   Rist SE, Assidqi K, Zamani NP, *et al.* Suspended micro-sized PVC particles impair the performance

and decrease survival in the Asian green mussel Perna viridis. Mar Pollut Bull 2016; 111(1-2): 213-20.
[http://dx.doi.org/10.1016/j.marpolbul.2016.07.006] [PMID: 27491368]

[29]   Oliveira M, Ribeiro A, Hylland K, Guilhermino L. Single and combined effects of microplastics and pyrene on juveniles (0+ group) of the common goby *Pomatoschistus microps* (Teleostei, Gobiidae). Ecol Indic 2013; 34: 641-7.
[http://dx.doi.org/10.1016/j.ecolind.2013.06.019]

[30]   Sussarellu R, Suquet M, Thomas Y, *et al.* Oyster reproduction is affected by exposure to polystyrene microplastics. Proc Natl Acad Sci USA 2016; 113(9): 2430-5.
[http://dx.doi.org/10.1073/pnas.1519019113] [PMID: 26831072]

[31]   Tallec K, Huvet A, Di Poi C, *et al.* Nanoplastics impaired oyster free living stages, gametes and embryos. Environ Pollut 2018; 242(Pt B): 1226-35.
[http://dx.doi.org/10.1016/j.envpol.2018.08.020] [PMID: 30118910]

[32]   Della Torre C, Bergami E, Salvati A, *et al.* Accumulation and embryotoxicity of polystyrene nanoparticles at early stage of development of sea urchin embryos *Paracentrotus lividus*. Environ Sci Technol 2014; 48(20): 12302-11.
[http://dx.doi.org/10.1021/es502569w] [PMID: 25260196]

[33]   Lei L, Wu S, Lu S, *et al.* Microplastic particles cause intestinal damage and other adverse effects in zebrafish *Danio rerio* and nematode *Caenorhabditis elegans*. Sci Total Environ 2018; 619-620: 1-8.
[http://dx.doi.org/10.1016/j.scitotenv.2017.11.103] [PMID: 29136530]

[34]   Wang Y, Zhang D, Zhang M, *et al.* Effects of ingested polystyrene microplastics on brine shrimp, *Artemia parthenogenetica*. Environ Pollut 2019; 244: 715-22.
[http://dx.doi.org/10.1016/j.envpol.2018.10.024] [PMID: 30384077]

[35]   Espinosa C, Cuesta A, Esteban MÁ. Effects of dietary polyvinylchloride microparticles on general health, immune status and expression of several genes related to stress in gilthead seabream (*Sparus aurata* L.). Fish Shellfish Immunol 2017; 68: 251-9.
[http://dx.doi.org/10.1016/j.fsi.2017.07.006] [PMID: 28684324]

[36]   Yu P, Liu Z, Wu D, Chen M, Lv W, Zhao Y. Accumulation of polystyrene microplastics in juvenile *Eriocheir sinensis* and oxidative stress effects in the liver. Aquat Toxicol 2018; 200: 28-36.
[http://dx.doi.org/10.1016/j.aquatox.2018.04.015] [PMID: 29709883]

[37]   Tang J, Ni X, Zhou Z, Wang L, Lin S. Acute microplastic exposure raises stress response and suppresses detoxification and immune capacities in the scleractinian coral *Pocillopora damicornis*. Environ Pollut 2018; 243(Pt A): 66-74.
[http://dx.doi.org/10.1016/j.envpol.2018.08.045] [PMID: 30172125]

[38]   Brandts I, Teles M, Gonçalves AP, *et al.* Effects of nanoplastics on *Mytilus galloprovincialis* after individual and combined exposure with carbamazepine. Sci Total Environ 2018; 643: 775-84. a
[http://dx.doi.org/10.1016/j.scitotenv.2018.06.257] [PMID: 29958167]

[39]   Brandts I, Teles M, Tvarijonaviciute A, *et al.* Effects of polymethylmethacrylate nanoplastics on Dicentrarchus labrax. Genomics 2018; 110(6): 435-41. b
[http://dx.doi.org/10.1016/j.ygeno.2018.10.006] [PMID: 30316739]

[40]   LeMoine CMR, Kelleher BM, Lagarde R, Northam C, Elebute OO, Cassone BJ. Transcriptional effects of polyethylene microplastics ingestion in developing zebrafish (*Danio rerio*). Environ Pollut 2018; 243(Pt A): 591-600.
[http://dx.doi.org/10.1016/j.envpol.2018.08.084] [PMID: 30218869]

[41]   Mazurais D, Ernande B, Quazuguel P, *et al.* Evaluation of the impact of polyethylene microbeads ingestion in European sea bass (Dicentrarchus labrax) larvae. Mar Environ Res 2015; 112(Pt A): 78-85.
[http://dx.doi.org/10.1016/j.marenvres.2015.09.009] [PMID: 26412109]

[42]   Au SY, Bruce TF, Bridges WC, Klaine SJ. Responses of *Hyalella azteca* to acute and chronic

microplastic exposures. Environ Toxicol Chem 2015; 34(11): 2564-72.
[http://dx.doi.org/10.1002/etc.3093] [PMID: 26042578]

[43]   Avio CG, Gorbi S, Milan M, *et al.* Pollutants bioavailability and toxicological risk from microplastics to marine mussels. Environ Pollut 2015; 198: 211-22.
[http://dx.doi.org/10.1016/j.envpol.2014.12.021] [PMID: 25637744]

[44]   Van Cauwenberghe L, Claessens M, Vandegehuchte MB, Janssen CR. Microplastics are taken up by mussels (Mytilus edulis) and lugworms (*Arenicola marina*) living in natural habitats. Environ Pollut 2015; 199: 10-7.
[http://dx.doi.org/10.1016/j.envpol.2015.01.008] [PMID: 25617854]

[45]   Kaposi KL, Mos B, Kelaher BP, Dworjanyn SA. Ingestion of microplastic has limited impact on a marine larva. Environ Sci Technol 2014; 48(3): 1638-45.
[http://dx.doi.org/10.1021/es404295e] [PMID: 24341789]

[46]   Nobre CR, Santana MFM, Maluf A, *et al.* Assessment of microplastic toxicity to embryonic development of the sea urchin Lytechinus variegatus (Echinodermata: Echinoidea). Mar Pollut Bull 2015; 92(1-2): 99-104.
[http://dx.doi.org/10.1016/j.marpolbul.2014.12.050] [PMID: 25662316]

[47]   Magni S, Gagné F, André C, *et al.* Evaluation of uptake and chronic toxicity of virgin polystyrene microbeads in freshwater zebra mussel Dreissena polymorpha (Mollusca: Bivalvia). Sci Total Environ 2018; 631-632: 778-88.
[http://dx.doi.org/10.1016/j.scitotenv.2018.03.075] [PMID: 29544181]

[48]   Veneman WJ, Spaink HP, Brun NR, Bosker T, Vijver MG. Pathway analysis of systemic transcriptome responses to injected polystyrene particles in zebrafish larvae. Aquat Toxicol 2017; 190: 112-20.
[http://dx.doi.org/10.1016/j.aquatox.2017.06.014] [PMID: 28704660]

[49]   Chen Q, Gundlach M, Yang S, *et al.* Quantitative investigation of the mechanisms of microplastics and nanoplastics toward zebrafish larvae locomotor activity. Sci Total Environ 2017; 584-585: 1022-31.
[http://dx.doi.org/10.1016/j.scitotenv.2017.01.156] [PMID: 28185727]

[50]   Oliveira M, Ribeiro A, Hylland K, Guilhermino L. Single and combined effects of microplastics and pyrene on juveniles (0+ group) of the common goby Pomatoschistus microps (Teleostei, Gobiidae). Ecol Indic 2013; 34: 641-7.
[http://dx.doi.org/10.1016/j.ecolind.2013.06.019]

[51]   Capolupo M, Franzellitti S, Valbonesi P, Lanzas CS, Fabbri E. Uptake and transcriptional effects of polystyrene microplastics in larval stages of the Mediterranean mussel Mytilus galloprovincialis. Environ Pollut 2018; 241: 1038-47.
[http://dx.doi.org/10.1016/j.envpol.2018.06.035] [PMID: 30029311]

[52]   Lassen C, Hansen SF, Magnusson K, *et al.* Microplastics: occurrence, effects and sources of releases to the environment in Denmark. Copenhagen: Danish Environmental Protection Agency 2015.

[53]   Dris R, Gasperi J, Saad M, Mirande C, Tassin B. Synthetic fibers in atmospheric fallout: A source of microplastics in the environment? Mar Pollut Bull 2016; 104(1-2): 290-3.
[http://dx.doi.org/10.1016/j.marpolbul.2016.01.006] [PMID: 26787549]

[54]   De Souza Machado AA, Lau CW, Till J, *et al.* Impacts of microplastics on the soil biophysical environment. Environ Sci Technol 2018; 52(17): 9656-65.
[http://dx.doi.org/10.1021/acs.est.8b02212] [PMID: 30053368]

[55]   Steinmetz Z, Wollmann C, Schaefer M, *et al.* Plastic mulching in agriculture. Trading short-term agronomic benefits for long-term soil degradation? Sci Total Environ 2016; 550: 690-705.
[http://dx.doi.org/10.1016/j.scitotenv.2016.01.153] [PMID: 26849333]

[56]   Zhao S, Zhu L, Li D. Microscopic anthropogenic litter in terrestrial birds from Shanghai, China: Not

only plastics but also natural fibers. Sci Total Environ 2016; 550: 1110-5.
[http://dx.doi.org/10.1016/j.scitotenv.2016.01.112] [PMID: 26874248]

[57]   Holland ER, Mallory ML, Shutler D. Plastics and other anthropogenic debris in freshwater birds from Canada. Sci Total Environ 2016; 571: 251-8.
[http://dx.doi.org/10.1016/j.scitotenv.2016.07.158] [PMID: 27476006]

[58]   Huerta Lwanga E, Mendoza Vega J, Ku Quej V, *et al.* Field evidence for transfer of plastic debris along a terrestrial food chain. Sci Rep 2017; 7(1): 14071.
[http://dx.doi.org/10.1038/s41598-017-14588-2] [PMID: 29074893]

[59]   Panebianco A, Nalbone L, Giarratana F, Ziino G. First discoveries of microplastics in terrestrial snails. Food Control 2019; 106: 106722.
[http://dx.doi.org/10.1016/j.foodcont.2019.106722]

[60]   Duis K, Coors A. Microplastics in the aquatic and terrestrial environment: sources (with a specific focus on personal care products), fate and effects. Environ Sci Eur 2016; 28(1): 2.
[http://dx.doi.org/10.1186/s12302-015-0069-y] [PMID: 27752437]

[61]   Mattsson K, Johnson EV, Malmendal A, Linse S, Hansson LA, Cedervall T. Brain damage and behavioural disorders in fish induced by plastic nanoparticles delivered through the food chain. Sci Rep 2017; 7(1): 11452.
[http://dx.doi.org/10.1038/s41598-017-10813-0] [PMID: 28904346]

[62]   Deng Y, Zhang Y, Lemos B, Ren H. Tissue accumulation of microplastics in mice and biomarker responses suggest widespread health risks of exposure. Sci Rep 2017; 7(1): 46687.
[http://dx.doi.org/10.1038/srep46687] [PMID: 28436478]

[63]   Hwang J, Choi D, Han S, Choi J, Hong J. An assessment of the toxicity of polypropylene microplastics in human derived cells. Sci Total Environ 2019; 684: 657-69.
[http://dx.doi.org/10.1016/j.scitotenv.2019.05.071] [PMID: 31158627]

[64]   Jiang X, Chang Y, Zhang T, Qiao Y, Klobučar G, Li M. Toxicological effects of polystyrene microplastics on earthworm (Eisenia fetida). Environ Pollut 2020; 259: 113896.
[http://dx.doi.org/10.1016/j.envpol.2019.113896] [PMID: 31918148]

[65]   Zhao S, Zhu L, Li D. Microscopic anthropogenic litter in terrestrial birds from Shanghai, China: Not only plastics but also natural fibers. Sci Total Environ 2016; 550: 1110-5.
[http://dx.doi.org/10.1016/j.scitotenv.2016.01.112] [PMID: 26874248]

[66]   Holland ER, Mallory ML, Shutler D. Plastics and other anthropogenic debris in freshwater birds from Canada. Sci Total Environ 2016; 571: 251-8.
[http://dx.doi.org/10.1016/j.scitotenv.2016.07.158] [PMID: 27476006]

[67]   Gil-Delgado JA, Guijarro D, Gosálvez RU, López-Iborra GM, Ponz A, Velasco A. Presence of plastic particles in waterbirds faeces collected in Spanish lakes. Environ Pollut 2017; 220(Pt A): 732-6.
[http://dx.doi.org/10.1016/j.envpol.2016.09.054] [PMID: 27667676]

[68]   Lwanga EH, Vega JM, Quej VK, *et al.* Field evidence for transfer of plastic debris along a terrestrial food chain. Sci Rep 2017; 7(1): 1-7.
[PMID: 28127051]

[69]   Hussain N, Jaitley V, Florence AT. Recent advances in the understanding of uptake of microparticulates across the gastrointestinal lymphatics. Adv Drug Deliv Rev 2001; 50(1-2): 107-42.
[http://dx.doi.org/10.1016/S0169-409X(01)00152-1] [PMID: 11489336]

[70]   He L, Wu D, Rong H, Li M, Tong M, Kim H. Influence of nano-and microplastic particles on the transport and deposition behaviors of bacteria in quartz sand. Environ Sci Technol 2018; 52(20): acs.est.8b01673.
[http://dx.doi.org/10.1021/acs.est.8b01673] [PMID: 30204419]

[71]   Ren X, Tang J, Liu X, Liu Q. Effects of microplastics on greenhouse gas emissions and the microbial community in fertilized soil. Environ Pollut 2020; 256: 113347.

[http://dx.doi.org/10.1016/j.envpol.2019.113347] [PMID: 31672352]

[72]   Parthasarathy A, Tyler A C, Hoffman M J, Savka M A, Hudson A O. Is plastic pollution in aquatic and terrestrial environments a driver for the transmission of pathogens and the evolution of antibiotic resistance? Environ Sci Technol 2019; 53(4): 1744-5.
[http://dx.doi.org/10.1021/acs.est.8b07287]

[73]   Viršek MK, Lovšin MN, Koren Š, Kržan A, Peterlin M. Microplastics as a vector for the transport of the bacterial fish pathogen species Aeromonas salmonicida. Mar Pollut Bull 2017; 125(1-2): 301-9.
[http://dx.doi.org/10.1016/j.marpolbul.2017.08.024] [PMID: 28889914]

[74]   Yokota K, Waterfield H, Hastings C, Davidson E, Kwietniewski E, Wells B. Finding the missing piece of the aquatic plastic pollution puzzle: Interaction between primary producers and microplastics. Limnol Oceanogr Lett 2017; 2(4): 91-104.
[http://dx.doi.org/10.1002/lol2.10040]

[75]   Zhang C, Chen X, Wang J, Tan L. Toxic effects of microplastic on marine microalgae Skeletonema costatum: Interactions between microplastic and algae. Environ Pollut 2017; 220(Pt B): 1282-8.
[http://dx.doi.org/10.1016/j.envpol.2016.11.005] [PMID: 27876228]

[76]   Awet TT, Kohl Y, Meier F, *et al.* Effects of polystyrene nanoparticles on the microbiota and functional diversity of enzymes in soil. Environ Sci Eur 2018; 30(1): 11.
[http://dx.doi.org/10.1186/s12302-018-0140-6] [PMID: 29963347]

[77]   Souza Machado AA, Kloas W, Zarfl C, Hempel S, Rillig MC. Microplastics as an emerging threat to terrestrial ecosystems. Glob Change Biol 2018; 24(4): 1405-16.
[http://dx.doi.org/10.1111/gcb.14020] [PMID: 29245177]

[78]   Liu H, Yang X, Liu G, *et al.* Response of soil dissolved organic matter to microplastic addition in Chinese loess soil. Chemosphere 2017; 185: 907-17.
[http://dx.doi.org/10.1016/j.chemosphere.2017.07.064] [PMID: 28747000]

[79]   Powell JR, Rillig MC. Biodiversity of arbuscular mycorrhizal fungi and ecosystem function. New Phytol 2018; 220(4): 1059-75.
[http://dx.doi.org/10.1111/nph.15119] [PMID: 29603232]

[80]   Wang F, Zhang X, Zhang S, Zhang S, Sun Y. Interactions of microplastics and cadmium on plant growth and arbuscular mycorrhizal fungal communities in an agricultural soil. Chemosphere 2020; 254: 126791.
[http://dx.doi.org/10.1016/j.chemosphere.2020.126791] [PMID: 32320834]

[81]   Chen H, Wang Y, Sun X, Peng Y, Xiao L. Mixing effect of polylactic acid microplastic and straw residue on soil property and ecological function. Chemosphere 2020; 243: 125271.
[http://dx.doi.org/10.1016/j.chemosphere.2019.125271] [PMID: 31760289]

[82]   Yang X, Bento CPM, Chen H, *et al.* Influence of microplastic addition on glyphosate decay and soil microbial activities in Chinese loess soil. Environ Pollut 2018; 242(Pt A): 338-47.
[http://dx.doi.org/10.1016/j.envpol.2018.07.006] [PMID: 29990941]

[83]   Fei Y, Huang S, Zhang H, *et al.* Response of soil enzyme activities and bacterial communities to the accumulation of microplastics in an acid cropped soil. Sci Total Environ 2020; 707: 135634.
[http://dx.doi.org/10.1016/j.scitotenv.2019.135634] [PMID: 31761364]

[84]   Chai B, Li X, Liu H, Lu G, Dang Z, Yin H. Bacterial communities on soil microplastic at Guiyu, an E-Waste dismantling zone of China. Ecotoxicol Environ Saf 2020; 195: 110521.
[http://dx.doi.org/10.1016/j.ecoenv.2020.110521] [PMID: 32222597]

[85]   Zhang M, Zhao Y, Qin X, *et al.* Microplastics from mulching film is a distinct habitat for bacteria in farmland soil. Sci Total Environ 2019; 688: 470-8.
[http://dx.doi.org/10.1016/j.scitotenv.2019.06.108] [PMID: 31254812]

[86]   Huang Y, Zhao Y, Wang J, *et al.* LDPE microplastic films alter microbial community composition and enzymatic activities in soil. Environ Pollut 2019; 254(Pt A): 112983.

[http://dx.doi.org/10.1016/j.envpol.2019.112983] [PMID: 31394342]

[87]   Yi M, Zhou S, Zhang L, Ding S. The effects of three different microplastics on enzyme activities and microbial communities in soil. Water Environ Res 2021; 93(1): 24-32.
[http://dx.doi.org/10.1002/wer.1327] [PMID: 32187766]

[88]   Bandmann V, Müller JD, Köhler T, Homann U. Uptake of fluorescent nano beads into BY2-cells involves clathrin-dependent and clathrin-independent endocytosis. FEBS Lett 2012; 586(20): 3626-32.
[http://dx.doi.org/10.1016/j.febslet.2012.08.008] [PMID: 23046971]

[89]   Li L, Zhou Q, Yin N, Tu C, Luo Y. Uptake and accumulation of microplastics in an edible plant. Kexue Tongbao 2019; 64(9): 928-34.
[http://dx.doi.org/10.1360/N972018-00845]

[90]   Qi Y, Yang X, Pelaez AM, *et al.* Macro- and micro- plastics in soil-plant system: Effects of plastic mulch film residues on wheat (Triticum aestivum) growth. Sci Total Environ 2018; 645: 1048-56.
[http://dx.doi.org/10.1016/j.scitotenv.2018.07.229] [PMID: 30248830]

[91]   Pohl M, Graf F, Buttler A, Rixen C. The relationship between plant species richness and soil aggregate stability can depend on disturbance. Plant Soil 2012; 355(1-2): 87-102.
[http://dx.doi.org/10.1007/s11104-011-1083-5]

[92]   Pérès G, Cluzeau D, Menasseri S, *et al.* Mechanisms linking plant community properties to soil aggregate stability in an experimental grassland plant diversity gradient. Plant Soil 2013; 373(1-2): 285-99.
[http://dx.doi.org/10.1007/s11104-013-1791-0]

[93]   Wan Y, Wu C, Xue Q, Hui X. Effects of plastic contamination on water evaporation and desiccation cracking in soil. Sci Total Environ 2019; 654: 576-82.
[http://dx.doi.org/10.1016/j.scitotenv.2018.11.123] [PMID: 30447596]

[94]   Wagg C, Bender SF, Widmer F, van der Heijden MGA. Soil biodiversity and soil community composition determine ecosystem multifunctionality. Proc Natl Acad Sci USA 2014; 111(14): 5266-70.
[http://dx.doi.org/10.1073/pnas.1320054111] [PMID: 24639507]

[95]   van der Heijden MGA, Bruin S, Luckerhoff L, van Logtestijn RSP, Schlaeppi K. A widespread plant-fungal-bacterial symbiosis promotes plant biodiversity, plant nutrition and seedling recruitment. ISME J 2016; 10(2): 389-99.
[http://dx.doi.org/10.1038/ismej.2015.120] [PMID: 26172208]

[96]   Qi Y, Ossowicki A, Yang X, *et al.* Effects of plastic mulch film residues on wheat rhizosphere and soil properties. J Hazard Mater 2020; 387: 121711.
[http://dx.doi.org/10.1016/j.jhazmat.2019.121711] [PMID: 31806445]

[97]   Rillig MC, Ryo M, Lehmann A, *et al.* The role of multiple global change factors in driving soil functions and microbial biodiversity. Science 2019; 366(6467): 886-90.
[http://dx.doi.org/10.1126/science.aay2832] [PMID: 31727838]

[98]   Wang W, Ge J, Yu X, Li H. Environmental fate and impacts of microplastics in soil ecosystems: Progress and perspective. Sci Total Environ 2020; 708: 134841.
[http://dx.doi.org/10.1016/j.scitotenv.2019.134841] [PMID: 31791759]

[99]   Souza Machado AA, Kloas W, Zarfl C, Hempel S, Rillig MC. Microplastics as an emerging threat to terrestrial ecosystems. Glob Change Biol 2018; 24(4): 1405-16.
[http://dx.doi.org/10.1111/gcb.14020] [PMID: 29245177]

[100]   Zhang GS, Liu YF. The distribution of microplastics in soil aggregate fractions in southwestern China. Sci Total Environ 2018; 642: 12-20.
[http://dx.doi.org/10.1016/j.scitotenv.2018.06.004] [PMID: 29894871]

[101]   Lozano YM, Rillig MC. Effects of microplastic fibers and drought on plant communities. Environ Sci Technol 2020; 54(10): 6166-73.

[http://dx.doi.org/10.1021/acs.est.0c01051] [PMID: 32289223]

[102] Boots B, Russell CW, Green DS. Effects of microplastics in soil ecosystems: above and below ground. Environ Sci Technol 2019; 53(19): 11496-506.
[http://dx.doi.org/10.1021/acs.est.9b03304] [PMID: 31509704]

[103] Bosker T, Bouwman LJ, Brun NR, Behrens P, Vijver MG. Microplastics accumulate on pores in seed capsule and delay germination and root growth of the terrestrial vascular plant Lepidium sativum. Chemosphere 2019; 226: 774-81.
[http://dx.doi.org/10.1016/j.chemosphere.2019.03.163] [PMID: 30965248]

[104] Jiang X, Chen H, Liao Y, Ye Z, Li M, Klobučar G. Ecotoxicity and genotoxicity of polystyrene microplastics on higher plant Vicia faba. Environ Pollut 2019; 250: 831-8.
[http://dx.doi.org/10.1016/j.envpol.2019.04.055] [PMID: 31051394]

[105] Li L, Luo Y, Peijnenburg WJGM, Li R, Yang J, Zhou Q. Confocal measurement of microplastics uptake by plants. MethodsX 2020; 7: 100750.
[http://dx.doi.org/10.1016/j.mex.2019.11.023] [PMID: 32021814]

[106] Li L, Zhou Q, Yin N, Tu C, Luo Y. Uptake and accumulation of microplastics in an edible plant. Kexue Tongbao 2019; 64(9): 928-34.
[http://dx.doi.org/10.1360/N972018-00845]

[107] Chae Y, An YJ. Nanoplastic ingestion induces behavioral disorders in terrestrial snails: trophic transfer effects *via* vascular plants. Environ Sci Nano 2020; 7(3): 975-83.
[http://dx.doi.org/10.1039/C9EN01335K]

[108] Lee H, Shim WJ, Kwon JH. Sorption capacity of plastic debris for hydrophobic organic chemicals. Sci Total Environ 2014; 470-471: 1545-52.
[http://dx.doi.org/10.1016/j.scitotenv.2013.08.023] [PMID: 24012321]

[109] Velzeboer I, Kwadijk CJAF, Koelmans AA. Strong sorption of PCBs to nanoplastics, microplastics, carbon nanotubes, and fullerenes. Environ Sci Technol 2014; 48(9): 4869-76.
[http://dx.doi.org/10.1021/es405721v] [PMID: 24689832]

[110] Antunes JC, Frias JGL, Micaelo AC, Sobral P. Resin pellets from beaches of the Portuguese coast and adsorbed persistent organic pollutants. Estuar Coast Shelf Sci 2013; 130: 62-9.
[http://dx.doi.org/10.1016/j.ecss.2013.06.016]

[111] Mai L, Bao LJ, Shi L, Liu LY, Zeng EY. Polycyclic aromatic hydrocarbons affiliated with microplastics in surface waters of Bohai and Huanghai Seas, China. Environ Pollut 2018; 241: 834-40.
[http://dx.doi.org/10.1016/j.envpol.2018.06.012] [PMID: 29909309]

[112] Romeo T, D'Alessandro M, Esposito V, *et al.* Environmental quality assessment of Grand Harbour (Valletta, Maltese Islands): a case study of a busy harbour in the Central Mediterranean Sea. Environ Monit Assess 2015; 187(12): 747.
[http://dx.doi.org/10.1007/s10661-015-4950-3] [PMID: 26563234]

[113] León VM, García I, González E, Samper R, Fernández-González V, Muniategui-Lorenzo S. Potential transfer of organic pollutants from littoral plastics debris to the marine environment. Environ Pollut 2018; 236: 442-53.
[http://dx.doi.org/10.1016/j.envpol.2018.01.114] [PMID: 29414369]

[114] Koelmans AA, Bakir A, Burton GA, Janssen CR. Microplastic as a vector for chemicals in the aquatic environment: critical review and model-supported reinterpretation of empirical studies. Environ Sci Technol 2016; 50(7): 3315-26.
[http://dx.doi.org/10.1021/acs.est.5b06069] [PMID: 26946978]

[115] Scopetani C, Cincinelli A, Martellini T, *et al.* Ingested microplastic as a two-way transporter for PBDEs in Talitrus saltator. Environ Res 2018; 167: 411-7.
[http://dx.doi.org/10.1016/j.envres.2018.07.030] [PMID: 30118960]

[116] Brennecke D, Duarte B, Paiva F, Caçador I, Canning-Clode J. Microplastics as vector for heavy metal

contamination from the marine environment. Estuar Coast Shelf Sci 2016; 178: 189-95.
[http://dx.doi.org/10.1016/j.ecss.2015.12.003]

[117]   Deanin RD. Additives in plastics. Environ Health Perspect 1975; 11: 35-9.
[http://dx.doi.org/10.1289/ehp.751135] [PMID: 1175566]

[118]   Hauser R, Calafat AM. Phthalates and human health. Occup Environ Med 2005; 62(11): 806-18.
[http://dx.doi.org/10.1136/oem.2004.017590] [PMID: 16234408]

[119]   Dobaradaran S, Schmidt TC, Nabipour I, *et al.* Characterization of plastic debris and association of
metals with microplastics in coastline sediment along the Persian Gulf. Waste Manag 2018; 78: 649-
58.
[http://dx.doi.org/10.1016/j.wasman.2018.06.037] [PMID: 32559956]

[120]   Zon NF, Iskendar A, Azman S, Sarijan S, Ismail R. Sorptive behaviour of chromium on polyethylene
microbeads in artificial seawater. MATEC Web of Conferences. Vol. 250: 06001.
[http://dx.doi.org/10.1051/matecconf/201825006001]

[121]   Vedolin MC, Teophilo CYS, Turra A, Figueira RCL. Spatial variability in the concentrations of metals
in beached microplastics. Mar Pollut Bull 2018; 129(2): 487-93.
[http://dx.doi.org/10.1016/j.marpolbul.2017.10.019] [PMID: 29033167]

[122]   Lee H, Shim WJ, Kwon JH. Sorption capacity of plastic debris for hydrophobic organic chemicals. Sci
Total Environ 2014; 470-471: 1545-52.
[http://dx.doi.org/10.1016/j.scitotenv.2013.08.023] [PMID: 24012321]

[123]   Hüffer T, Hofmann T. Sorption of non-polar organic compounds by micro-sized plastic particles in
aqueous solution. Environ Pollut 2016; 214: 194-201.
[http://dx.doi.org/10.1016/j.envpol.2016.04.018] [PMID: 27086075]

[124]   Wu P, Cai Z, Jin H, Tang Y. Adsorption mechanisms of five bisphenol analogues on PVC
microplastics. Sci Total Environ 2019; 650(Pt 1): 671-8.
[http://dx.doi.org/10.1016/j.scitotenv.2018.09.049] [PMID: 30212696]

[125]   Seidensticker S, Grathwohl P, Lamprecht J, Zarfl C. A combined experimental and modeling study to
evaluate pH-dependent sorption of polar and non-polar compounds to polyethylene and polystyrene
microplastics. Environ Sci Eur 2018; 30(1): 30.
[http://dx.doi.org/10.1186/s12302-018-0155-z] [PMID: 30148026]

[126]   Guo X, Pang J, Chen S, Jia H. Sorption properties of tylosin on four different microplastics.
Chemosphere 2018; 209: 240-5.
[http://dx.doi.org/10.1016/j.chemosphere.2018.06.100] [PMID: 29933160]

[127]   Liu F, Xu B, He Y, Brookes PC, Tang C, Xu J. Differences in transport behavior of natural soil
colloids of contrasting sizes from nanometer to micron and the environmental implications. Sci Total
Environ 2018; 634: 802-10.
[http://dx.doi.org/10.1016/j.scitotenv.2018.03.381] [PMID: 29660881]

[128]   Xu B, Liu F, Brookes PC, Xu J. Microplastics play a minor role in tetracycline sorption in the presence
of dissolved organic matter. Environ Pollut 2018; 240: 87-94.
[http://dx.doi.org/10.1016/j.envpol.2018.04.113] [PMID: 29729573]

[129]   Zhang GS, Liu YF. The distribution of microplastics in soil aggregate fractions in southwestern China.
Sci Total Environ 2018; 642: 12-20.
[http://dx.doi.org/10.1016/j.scitotenv.2018.06.004] [PMID: 29894871]

[130]   Seidensticker S, Zarfl C, Cirpka OA, Fellenberg G, Grathwohl P. Shift in mass transfer of wastewater
contaminants from microplastics in the presence of dissolved substances. Environ Sci Technol 2017;
51(21): 12254-63.
[http://dx.doi.org/10.1021/acs.est.7b02664] [PMID: 28965391]

[131]   Shen XC, Li DC, Sima XF, Cheng HY, Jiang H. The effects of environmental conditions on the
enrichment of antibiotics on microplastics in simulated natural water column. Environ Res 2018; 166:

377-83.
[http://dx.doi.org/10.1016/j.envres.2018.06.034] [PMID: 29935450]

[132] Wu C, Zhang K, Huang X, Liu J. Sorption of pharmaceuticals and personal care products to polyethylene debris. Environ Sci Pollut Res Int 2016; 23(9): 8819-26.
[http://dx.doi.org/10.1007/s11356-016-6121-7] [PMID: 26810664]

[133] Chen W, Ouyang ZY, Qian C, Yu HQ. Induced structural changes of humic acid by exposure of polystyrene microplastics: A spectroscopic insight. Environ Pollut 2018; 233: 1-7.
[http://dx.doi.org/10.1016/j.envpol.2017.10.027] [PMID: 29049941]

[134] Adjei D, Sharma B, Labhasetwar V. Nanoparticles: cellular uptake and cytotoxicityAdv Exp Med Biol 2014; 811: 73-91.
[http://dx.doi.org/10.1007/978-94-017-8739-0_5] [PMID: 24683028]

[135] Heskett M, Takada H, Yamashita R, *et al.* Measurement of persistent organic pollutants (POPs) in plastic resin pellets from remote islands: Toward establishment of background concentrations for International Pellet Watch. Mar Pollut Bull 2012; 64(2): 445-8.
[http://dx.doi.org/10.1016/j.marpolbul.2011.11.004] [PMID: 22137935]

[136] Yeo BG, Takada H, Hosoda J, *et al.* Polycyclic aromatic hydrocarbons (PAHs) and hopanes in plastic resin pellets as markers of oil pollution *via* international pellet watch monitoring. Arch Environ Contam Toxicol 2017; 73(2): 196-206.
[http://dx.doi.org/10.1007/s00244-017-0423-8] [PMID: 28710501]

[137] Vallejo M, Fresnedo San Román M, Ortiz I, Irabien A. Overview of the PCDD/Fs degradation potential and formation risk in the application of advanced oxidation processes (AOPs) to wastewater treatment. Chemosphere 2015; 118: 44-56.
[http://dx.doi.org/10.1016/j.chemosphere.2014.05.077] [PMID: 24974140]

[138] Syberg K, Khan FR, Selck H, *et al.* Microplastics: addressing ecological risk through lessons learned. Environ Toxicol Chem 2015; 34(5): 945-53.
[http://dx.doi.org/10.1002/etc.2914] [PMID: 25655822]

[139] Weber A, Jeckel N, Wagner M. Combined effects of polystyrene microplastics and thermal stress on the freshwater mussel Dreissena polymorpha. Sci Total Environ 2020; 718: 137253.
[http://dx.doi.org/10.1016/j.scitotenv.2020.137253] [PMID: 32087582]

[140] Serra T, Barcelona A, Pous N, Salvadó V, Colomer J. Synergistic effects of water temperature, microplastics and ammonium as second and third order stressors on *Daphnia magna*. Environ Pollut 2020; 267: 115439.
[http://dx.doi.org/10.1016/j.envpol.2020.115439] [PMID: 32892007]

[141] Kim D, Chae Y, An YJ. Mixture toxicity of nickel and microplastics with different functional groups on *Daphnia magna*. Environ Sci Technol 2017; 51(21): 12852-8.
[http://dx.doi.org/10.1021/acs.est.7b03732] [PMID: 29019667]

[142] Axworthy JB, Padilla-Gamiño JL. Microplastics ingestion and heterotrophy in thermally stressed corals. Sci Rep 2019; 9(1): 18193.
[http://dx.doi.org/10.1038/s41598-019-54698-7] [PMID: 31796829]

[143] Kratina P, Watts TJ, Green DS, Kordas RL, O'Gorman EJ. Interactive effects of warming and microplastics on metabolism but not feeding rates of a key freshwater detritivore. Environ Pollut 2019; 255(Pt 2): 113259.
[http://dx.doi.org/10.1016/j.envpol.2019.113259] [PMID: 31563782]

[144] Jaikumar G, Baas J, Brun NR, Vijver MG, Bosker T. Acute sensitivity of three Cladoceran species to different types of microplastics in combination with thermal stress. Environ Pollut 2018; 239: 733-40.
[http://dx.doi.org/10.1016/j.envpol.2018.04.069] [PMID: 29723823]

[145] Corinaldesi C, Canensi S, Dell'Anno A, *et al.* Multiple impacts of microplastics can threaten marine habitat-forming species. Commun Biol 2021; 4(1): 431.

[http://dx.doi.org/10.1038/s42003-021-01961-1] [PMID: 33785849]

[146] Zhang P, Yan Z, Lu G, Ji Y. Single and combined effects of microplastics and roxithromycin on *Daphnia magna*. Environ Sci Pollut Res Int 2019; 26(17): 17010-20.
[http://dx.doi.org/10.1007/s11356-019-05031-2] [PMID: 30972681]

[147] Qu H, Ma R, Wang B, Yang J, Duan L, Yu G. Enantiospecific toxicity, distribution and bioaccumulation of chiral antidepressant venlafaxine and its metabolite in loach (Misgurnus anguillicaudatus) co-exposed to microplastic and the drugs. J Hazard Mater 2019; 370: 203-11.
[http://dx.doi.org/10.1016/j.jhazmat.2018.04.041] [PMID: 29706475]

[148] Fonte E, Ferreira P, Guilhermino L. Temperature rise and microplastics interact with the toxicity of the antibiotic cefalexin to juveniles of the common goby (*Pomatoschistus microps*): Post-exposure predatory behaviour, acetylcholinesterase activity and lipid peroxidation. Aquat Toxicol 2016; 180: 173-85.
[http://dx.doi.org/10.1016/j.aquatox.2016.09.015] [PMID: 27721112]

[149] Ward JE, Kach DJ. Marine aggregates facilitate ingestion of nanoparticles by suspension-feeding bivalves. Mar Environ Res 2009; 68(3): 137-42.
[http://dx.doi.org/10.1016/j.marenvres.2009.05.002] [PMID: 19525006]

[150] Chen CS, Anaya JM, Zhang S, *et al.* Effects of engineered nanoparticles on the assembly of exopolymeric substances from phytoplankton. PLoS One 2011; 6(7): e21865.
[http://dx.doi.org/10.1371/journal.pone.0021865] [PMID: 21811550]

[151] Wegner A, Besseling E, Foekema EM, Kamermans P, Koelmans AA. Effects of nanopolystyrene on the feeding behavior of the blue mussel (*Mytilus edulis* L.). Environ Toxicol Chem 2012; 31(11): 2490-7.
[http://dx.doi.org/10.1002/etc.1984] [PMID: 22893562]

[152] Della Torre C, Bergami E, Salvati A, *et al.* Accumulation and embryotoxicity of polystyrene nanoparticles at early stage of development of sea urchin embryos *Paracentrotus lividus*. Environ Sci Technol 2014; 48(20): 12302-11.
[http://dx.doi.org/10.1021/es502569w] [PMID: 25260196]

[153] Canesi L, Ciacci C, Bergami E, *et al.* Evidence for immunomodulation and apoptotic processes induced by cationic polystyrene nanoparticles in the hemocytes of the marine bivalve Mytilus. Mar Environ Res 2015; 111: 34-40.
[http://dx.doi.org/10.1016/j.marenvres.2015.06.008] [PMID: 26115607]

[154] Canesi L, Ciacci C, Fabbri R, *et al.* Interactions of cationic polystyrene nanoparticles with marine bivalve hemocytes in a physiological environment: Role of soluble hemolymph proteins. Environ Res 2016; 150: 73-81.
[http://dx.doi.org/10.1016/j.envres.2016.05.045] [PMID: 27257827]

[155] Bergami E, Bocci E, Vannuccini ML, *et al.* Nano-sized polystyrene affects feeding, behavior and physiology of brine shrimp Artemia franciscana larvae. Ecotoxicol Environ Saf 2016; 123: 18-25.
[http://dx.doi.org/10.1016/j.ecoenv.2015.09.021] [PMID: 26422775]

[156] Lee KW, Shim WJ, Kwon OY, Kang JH. Size-dependent effects of micro polystyrene particles in the marine copepod Tigriopus japonicus. Environ Sci Technol 2013; 47(19): 11278-83.
[http://dx.doi.org/10.1021/es401932b] [PMID: 23988225]

[157] Cole M, Galloway TS. Ingestion of nanoplastics and microplastics by Pacific oyster larvae. Environ Sci Technol 2015; 49(24): 14625-32.
[http://dx.doi.org/10.1021/acs.est.5b04099] [PMID: 26580574]

[158] Jeong CB, Won EJ, Kang HM, *et al.* Microplastic size-dependent toxicity, oxidative stress induction, and p-JNK and p-p38 activation in the monogonont rotifer (Brachionus koreanus). Environ Sci Technol 2016; 50(16): 8849-57.
[http://dx.doi.org/10.1021/acs.est.6b01441] [PMID: 27438693]

[159] Manabe M, Tatarazako N, Kinoshita M. Uptake, excretion and toxicity of nano-sized latex particles on medaka (*Oryzias latipes*) embryos and larvae. Aquat Toxicol 2011; 105(3-4): 576-81.
[http://dx.doi.org/10.1016/j.aquatox.2011.08.020] [PMID: 21946167]

[160] Lu Y, Zhang Y, Deng Y, *et al.* Uptake and accumulation of polystyrene microplastics in zebrafish (*Danio rerio*) and toxic effects in liver. Environ Sci Technol 2016; 50(7): 4054-60.
[http://dx.doi.org/10.1021/acs.est.6b00183] [PMID: 26950772]

[161] Sjollema SB, Redondo-Hasselerharm P, Leslie HA, Kraak MHS, Vethaak AD. Do plastic particles affect microalgal photosynthesis and growth? Aquat Toxicol 2016; 170: 259-61.
[http://dx.doi.org/10.1016/j.aquatox.2015.12.002] [PMID: 26675372]

[162] Belzagui F, Crespi M, Álvarez A, Gutiérrez-Bouzán C, Vilaseca M. Microplastics' emissions: Microfibers' detachment from textile garments. Environ Pollut 2019; 248: 1028-35.
[http://dx.doi.org/10.1016/j.envpol.2019.02.059] [PMID: 31091635]

[163] Manzoor S, Naqash N, Rashid G, Singh R. Plastic material degradation and formation of microplastic in the environment: A review. Mater Today Proc 2021.

[164] Conley K, Clum A, Deepe J, Lane H, Beckingham B. Wastewater treatment plants as a source of microplastics to an urban estuary: Removal efficiencies and loading per capita over one year. Water Res X 2019; 3: 100030.
[http://dx.doi.org/10.1016/j.wroa.2019.100030] [PMID: 31194047]

[165] Manzoor S, Kaur H, Singh R. Existence of Microplastic as Pollutant in Harike Wetland: An Analysis of Plastic Composition and First Report on Ramsar Wetland of India.

[166] Corradini F, Meza P, Eguiluz R, Casado F, Huerta-Lwanga E, Geissen V. Evidence of microplastic accumulation in agricultural soils from sewage sludge disposal. Sci Total Environ 2019; 671: 411-20.
[http://dx.doi.org/10.1016/j.scitotenv.2019.03.368] [PMID: 30933797]

[167] Yousuf A, Naseer M, Naqash N, Singh R. Isolation and identification of microplastic particles from agricultural soil and its detection by fluorescence microscope technique. Think India Journal 2019; 22(16): 3934-49.

[168] Gündoğdu S, Çevik C, Güzel E, Kilercioğlu S. Microplastics in municipal wastewater treatment plants in Turkey: a comparison of the influent and secondary effluent concentrations. Environ Monit Assess 2018; 190(11): 626.
[http://dx.doi.org/10.1007/s10661-018-7010-y] [PMID: 30280276]

[169] Manzoor S, Kaur H, Singh R. Analysis Of Nylon 6 As Microplastic In Harike Wetland By Comparing Its IR Spectra With Virgin Nylon 6 And 6.6. Eur J Mol Clin Med 2020; 7(07): 2020.

[170] Obbard RW. Microplastics in Polar Regions: The role of long range transport. Curr Opin Environ Sci Health 2018; 1: 24-9.
[http://dx.doi.org/10.1016/j.coesh.2017.10.004]

[171] Naqash N, Prakash S, Kapoor D, Singh R. Interaction of freshwater microplastics with biota and heavy metals: a review. Environ Chem Lett 2020; 18(6): 1813-24.
[http://dx.doi.org/10.1007/s10311-020-01044-3]

[172] Woodall LC, Sanchez-Vidal A, Canals M, *et al.* The deep sea is a major sink for microplastic debris. R Soc Open Sci 2014; 1(4): 140317.
[http://dx.doi.org/10.1098/rsos.140317] [PMID: 26064573]

CHAPTER 8

Bioplastic as an Alternative to Microplastic

Rohan Samir Kumar Sachan[1], Manpreet Kaur Somal[2], Ritu Bala[1], Khushboo[1], Mukesh Kumar[1], Inderpal Devgon[1] and Arun Karnwal[1,*]

[1] *Department of Microbiology, School of Bioengineering and Biosciences, Lovely Professional University, Phagwara-144411, Punjab, India*

[2] *Department of Biotechnology, School of Bioengineering and Biosciences, Lovely Professional University, Phagwara-144411, Punjab, India*

Abstract: Microplastics pose an imminent risk to the marine environment, biota, and ecosystem. Their consumption threatens organisms because of the material's ability to absorb and concentrate environmental contaminants in oceans and then transfer them through food chains. Microplastic may harm soil biota, such as earthworms, and can alter soil biophysical parameters, such as soil bulk density, aggregation, and water-holding capacity. To find alternatives to microplastics, scientists have developed biodegradable plastics that can be discarded in the environment and broken down quickly by the enzymatic activity of micro-organisms. Bioplastics are made from biological or renewable components. The bioplastic produced from potato peels, corn, sugarcane, wheat, rice, banana peels, and other natural materials is eco-friendly and biodegradable. Bioplastic is also known as Low-carbon plastic. The use of low-carbon plastic aids in the regulation of global temperature rise. It is used to make toys, home interiors, shopping bags, bottles, labels, trash bags, and packaging materials. It has wide applications for bone nails and tissue scaffolds in the medical industry. Its development also faces other obstacles, including price difficulties, technical improvements, and waste collection and treatment. Synthesis and characterization methods will help overcome these obstacles. The present chapter will focus on bioplastic and its types, the synthesis of bioplastic, the difference between microplastic and bioplastic, and bioplastic as an alternative approach.

Keywords: Additives, Anthropogenic contaminants, Biocompatible, Biodegradable, Bioplastic, Biopolymer, Economic feasibility, Human health, Microplastic human, Microplastic ingestion, Multi-stress, Nano-plastics, PHA-based composites, PHA-producing bacteria, Plastic pollution, Plastic, Polyhydroxyalkonate, Polyhydroxy butyrate, Polystyrene microplastic, Waste feedstocks.

* **Corresponding author Arun Karnwal:** Department of Microbiology, School of Bioengineering and Biosciences, Lovely Professional University, Phagwara-144411, Punjab, India; E-mail: arunkarnwal@gmail.com

Rahul Singh and Neeta Raj Sharma (Eds.)

INTRODUCTION

The monomers' polymerization extracted from the hydrocarbons creates synthetic organic polymers in nature. In 2016, the annual production of plastics reached 335 million worldwide [1]. The pollutants of plastics enter the ocean in many ways and constantly gather in the marine environment affecting marine life. As plastics are a prevalent and persistent pollutant to this date, they cause many impacts such as the effect on reproduction, ingestion, and non-native species translocation [2]. Land-based sources of plastics such as industries, riverine, urbanization, and tourism are more than sea-based plastic sources [3]. It indicates that more significant coastal site produces considerable plastic pollution [4]. The primary microplastics are microscopic; examples are resin pellets preproduction, blasting, toothpaste, microbeads in cosmetics, and media in drug delivery [5].

The degradation of larger plastics due to mechanical fragmentation and biological degradation leads to the formation of secondary microplastics [6]. The secondary microplastics include fabric from microfibers, fragments of plastic, and tire wire debris [7]. Microplastics are primarily available in marine organisms due to their spatial distribution and small size. For instance, microplastic found in shellfish, marine food, canned food, and mineral drinking water is stated to ingest by humans [8, 9]. From the coast of Rapa Nui, 80% of the *Decapterus muroadsi* sample has ingested microplastics. It has also been reported that 26 different species of fish from the various habitats of the red seacoast of Saudi Arabia also consumed the microplastic.

Renewable resources produce bioplastics as an alternative to microplastics or petrochemical-derived plastics [10]. Bioplastic is generally derived from natural resources. Bioplastic is categorized into two parts; biodegradable and non – biodegradable. Biodegradable plastic includes PLA (polylactic acid), PHA (polyhydroxyalkanoates), starch, and cellulose. Some plastics, including oil-based and biodegradable biobased plastic, can be incinerated as they are not entirely recycled and cause contamination in the recycling process. The non-biodegradable bioplastic includes bio-polyethylene terephthalate. The optically active PHAs are biodegradable polymers synthesized by bacteria during nutrient deprivation such as nitrogen, phosphate, and excess carbon. The microbial cell lysis and the microbial fermentation process are used for the PHAs production as they are biocompatible, biodegradable, and non-toxic polyesters. The diversity in their monomeric compositions is due to the variations in their physical properties as they carry wide applications [11].

HARMFUL EFFECT OF MICRO-PLASTIC

Today, micro-plastic is a top problem for the whole world; bio-plastic consumption shows many adverse effects on human and marine life. According to the Central Pollution Control Board, New Delhi, in 2014, 5.6 million tons of plastic were consumed in India, whose disposal is significant. Removal of industrial toxins and plastic waste in water bodies leads to an increase in the micro-plastic consumed by sea fish and seafood transfers that bioplastic to humans, which can act as a vector for different industrial toxins to enter the human body [12].

These insoluble synthetic solid particles cause pollution and become a high risk to marine life and humans, exposing micro-plastics to organisms [13].

Microplastics' physical and chemical effects are harming human and marine life see Fig. (1). In a 2016 UN report, more than 800 animal species engulfed or consumed micro-plastic. Among them, 220 species ingested micro-plastics get bloated in the organism's tract. As a part of the food chain, consumers may have consumed micro-plastic showing adverse effects like hemocyte aggregation and reduced respiratory function. It also activates an immune response in blue mussels and forms granulomas hepatic stress. In Japanese medaka, translocation of micro-plastic to the lymphatic system shows adverse effects on the immune system and cells. Even low dose consumption of chemicals applied to microplastic affects the biological system posing a risk to young humans and animals [14]. Adverse effects like breast cancer, defects in calcium metabolisms, cell apoptosis, female metal-estrogen, cell genotoxicity, porous bones (Osteoporoses), and cell conversion to cancerous cells are seen in humans [13, 15].

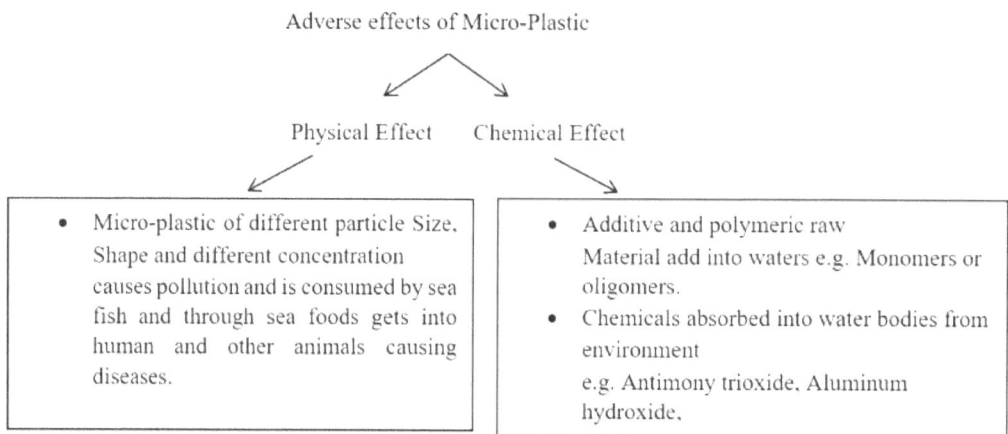

Fig. (1). Flow chart: Different effects of Micro-plastic.

However, marine organisms like fish ingest microplastic through biomagnification directly affects the consumers' health like humans and animals. Statistically, around 25% of marine fish are used as fishmeals and fish oil for human consumption, and it is evident that these contain a small proportion of microplastics [16].

BIOPLASTIC AS AN ALTERNATIVE

The disposal of synthetic plastics causes environmental problems and has prompted a hunt for alternatives. Bioplastics, both functionally and environmentally equivalent to synthetic plastics, are being offered as promising new materials to resolve these concerns. Natural polymeric components such as starch, cellulose, lignin, vegetable oil, and animal-derived compounds like proteins and lipids are used for bioplastics synthesis [17].

Plant-based Bioplastics

They are derived from potatoes or maize, rice, *etc*. Nicotiana, Elaeis, Brassica, tobacco, Gossypium, and Medicago are some plants that synthesize a range of PHAs see Fig. (**2**) [18]. Cellulose-based plastics produce from wood pulp to make wrappers. Starch from plants, most frequently corn or sugarcane, is fermented into lactic acid and subsequently polymerized. PLA is biodegradable and biocompatible in contact and is widely used for biomedical applications such as implants, sutures, and medication encapsulation [19]. Polyesters made from raw vegetable materials fermented with various bacterial strains are PHAs. They can be used in injection moulding to form vehicle parts and a variety of other applications [20]. Studies have shown that bioplastic derived from potato and yarn tubers possess good thermal, and mechanical properties and high biodegradability, *i.e.*, 43% and 26%, respectively [21].

Research has shown that the bioplastic produced from rice straw has good mechanical properties, with tensile strength and elongation at 45 MPa and 6.1% for dried and wet dumbbells, respectively, and 10 MPa and 63% for wet dumbbells and is degraded after 105 days [22]. Various fruits such as Solari's grape pomace, Banana pulp, peel, Apricot pomace, Pineapple peel waste, *etc.*, have produced PHB. 71% biodegradability in moist soil is shown by bioplastic prepared from potato peel waste [23]. A genetic construct designed to transform tobacco (*Nicotiana tabacum*) produces a renewable and biodegradable material called polyhydroxy butyrate (PHB). Plants could grow up to 18.8% dry-weight PHB in leaf tissue samples after transformation [24].

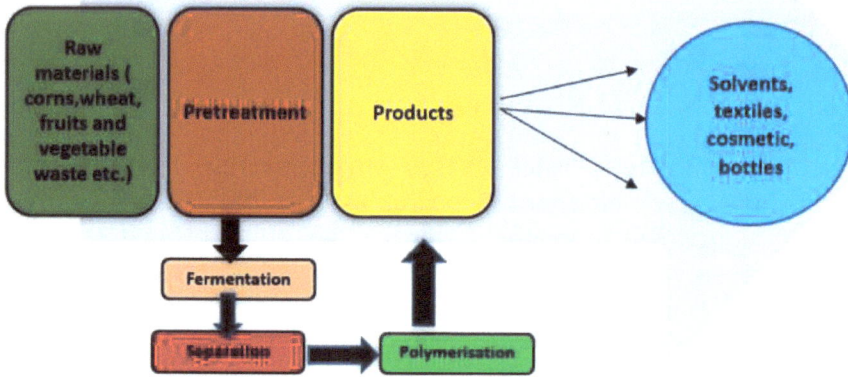

Fig. (2). Plants Based Bio-plastics.

Microorganism-based Bioplastic

Bacterial polyesters are a subgroup of bio-based polymers that polymerize by microbes. PHAs, for example, can be prepared enzymatically *in vivo* by microorganisms as a genuine natural polyester for intracellular storage. In contrast, PLA is made by fermenting lactic acid and then chemically polymerizing it. Only PLA and PHAs are entirely biodegradable and biodegradable in some form among bio-based polymers see Fig. (**3**) [25].

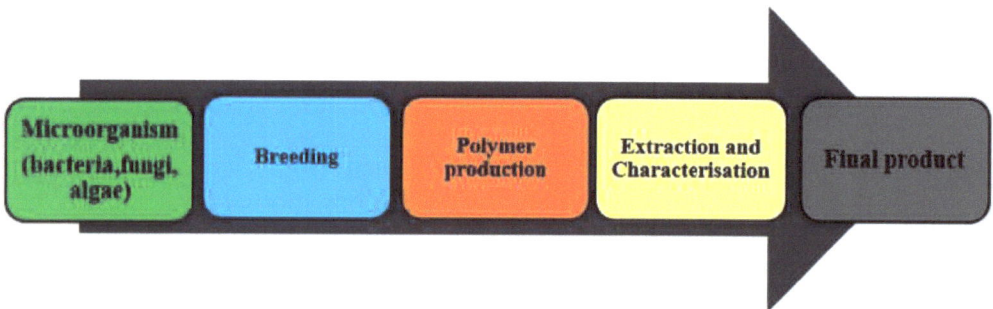

Fig. (3). Process of microorganism-based bioplastics.

Poly-3-hydroxybutyrate and polyhydroxyalkanoates are the widely used microbial bioplastic (Table **1**). Cyanobacteria blooms produce bioplastics by photosynthesis. Research has improved the cyanobacteria strain to produce plastics using self-synthesized glucose. It converts glucose into acetyl-CoA, which is further converted into acetoacetyl-CoA, followed by β-hydroxybutyric-CoA, and finally, the end product PHB.

Table 1. Micro-organisms and their derived bioplastic.

Microorganism	Bioplastic	References
Poly-3-hydroxybutyrate	*Pseudodonghicola xiamenensis*	[26]
Polyhydroxyalkanoates	*Halomonas hydrothermalis*	[27]
Polyhydroxybutyrate	*Bacillus megaterium*	[28]
Polyhydroxyalkanoates	*Thermus thermophiles*	[29]
Polyhydroxyalkanoates	*Cupriavidus*	[30]
Polyhydroxybutyrate	*Microcystis aeruginosa*	[31]
Polyhydroxybutyrate	*Caulobacter segnis*	[32]
pCB Poly (3-hydroxybutyrate-co-3-hydroxyhexanoate	*Engineered Ralstonia eutropha Re2133/*	[33]
Polyhydroxyalkanoates	*Scenedesmus sp*	[34]

PHAs have also been converted into fibres for use in non-woven fabrics. Food packaging and hygiene products like diapers, fishing nets, *etc.*, are examples of biodegradable packaging materials made with PHAs. PHAs have a wide range of uses in pharmacy and medicine. Bioplastics have exceptional biodegradability, which can assist the globe to deal with the growing litter problem, especially in rivers and seas. Plant-based bioplastics are durable and can be recycled, resulting in an economic drive to expand the bioplastic sector and give a better alternative for future environmental development [35].

MICROPLASTICS VS BIOPLASTICS

According to the latest studies, plastic manufacturing and consumption were predicted to be 2 million metric tons in the 1950s, which increased by about 396 million in 2016. Let's look at what plastic is and the different varieties it has. Let's also consider whether there is such a thing as "good" plastic.

Microplastics as Bane

Microplastics are highly minute particles of plastic that are generally less than 5mm in diameter [36]. Primary and secondary are the two main categories of microplastics. Primary microplastics are often known as microbeads. These polymers have been designed specifically for that size [37]. They're found in a variety of personal care items. Some of these microbeads are located in the things we utilize daily, such as sunscreen, body lotion, make-up, toothpaste, and scrubs.

The Plastic Soup Foundation launched its Beat the Microbead campaign in 2012. This initiative now has the backing of 95 non-governmental organizations from 40 countries across the world. Some nations, including the United States, Australia, and Canada, have banned microbeads due to this effort. However, this is only the beginning. The air blasting method also uses microplastics [38]. These microbeads remove paint or rust from various equipment, engines, and ships under high pressure. It's assumed that the compounds from these heavy metals infect these microbeads. This toxic mixture eventually ends up in our seas [39].

Microplastics that arise from more oversized garbage, such as plastic bags or bottles, are secondary. This garbage is primarily the result of poor waste management and ends up in the environment and the seas [39, 40]. Weather variables like tides, storms, temperature, light, or other physical stress cause the degradation of these wastes. In recent years, plastic's impact has been more challenging to detect. Microplastics do not degrade and are impossible to remove once they enter the environment.

Beer, tap water, sea salt, seafood, and other products contain microplastics. Greenpeace and the University of Incheon in South Korea found that 90 per cent of the sea salt they tested made in Asia had microplastics. A total of 39 different salt brands from throughout the world were investigated. Only three were free of microplastic [38].

Bioplastics

Waste substrates like corn, cassava, sugar beet, and sugar cane are natural renewable sources of bioplastic. PLAs and PHAs are the two most common forms of bioplastic. PLA stands for polylactic acid, produced by obtaining sugar from the plants and then fermenting it under controlled conditions [41]. Polyhydroxyalkanoates (PHA) are microorganism-engineered PHA. These bioplastics resemble standard plastic in appearance and behaviour, yet they are superior. Traditional plastic requires two-thirds the amount of energy that bioplastic does. Bioplastics, in general, are also biodegradable. They decompose into natural components that mix with the soil without harming it. These are all advantages of this environmentally friendly material [42].

Bioplastic has its own set of drawbacks. Some bioplastics, for instance, may not fully decompose quickly and leave hazardous residues or plastic particles behind [43]. Most bioplastics require a lot of heat to rot in a reasonable period. Bioplastic must be composted using a commercial composting method with a high temperature of over 58°C and relative humidity of 50% [44]. As you may assume, doing this at home is nearly difficult. Table **2** describes the comparison between bioplastics and micro-plastics.

Table 2. Comparison between Micro-plastics and Bioplastics.

-	Microplastics	Bioplastics
Definition	Microplastics are non-biodegradable tiny fragments of plastic with a length of less than 5 mm (0.2 inches) found in the environment due to plastic pollution.	Bioplastics are biodegradable materials derived from renewable resources that can address the problem of plastic trash strangling the earth and polluting the ecosystem.
Types	**Primary Microplastics** Small particles are discharged directly into the environment. Microplastics in the waters are estimated to account for 15-31% of all microplastics. **Secondary Microplastics** The source is decomposing more oversized plastic objects, such as plastic bags, bottles, or fishing nets. Microplastics make up 69-81 percent of all microplastics detected in the ocean.	Bioplastics made from maize starch are starch-based, and frequently combined with biodegradable polyesters. **Cellulose-based bioplastics** are made from esters and cellulose derivatives. **Protein-based bioplastics** are made from wheat gluten, casein, and milk.
Sources	They arise from synthetic clothing, laundry, and tyre abrasion from driving; microplastics were purposefully introduced to personal care items, such as microbeads in face scrubs.	They come from renewable organic resources, such as maize starch, milk, tapioca, and vegetable fats, among other biomass sources.
Characteristics	Microplastics are polymer chains of carbon and hydrogen atoms. Other microplastic compounds are phthalates, polybrominated diphenyl ethers, and tetrabromobisphenol A. Many of these chemical additions are separated from the plastics when released into the environment.	Bio-plastics are durable polymers, such as bio-based PE or PET, with the same qualities as their conventional counterparts. Although these bioplastics are theoretically identical to their fossil equivalents, they can lower a product's carbon impact. They can mechanically be recycled in the appropriate recycling processes.
Example	Polyester (PES) and acrylic (AC), polyamide (PA, *e.g.*, nylon), polypropylene (PP), polystyrene (PS), polyethylene (PE),	Bio-PE, Bio-PET, PLA, PHA, *etc.*

Furthermore, only a few cities have the necessary infrastructure to deal with this. On the other hand, bioplastic is commonly found in landfills due to poor waste management on our seas. They behave like petroleum-based plastic [45].

APPLICATIONS AND BIODEGRADATION OF BIOPLASTIC IN THE ENVIRONMENT

Due to the hazardous effect of petroleum-based plastic on nature, the potential applications of bioplastic are coming into consideration as an alternative in different areas such as packaging material, medical equipment, and coating materials. These biopolymers range from soft and smooth to brittle and rigid adhesives and elastomers, conditional to their monomer composition [46]. Bioplastic has advantageous properties to petroleum-based plastic, such as water-soluble, biocompatible, has zero toxicity to nature, and is biodegradable [47]. Such properties have expanded their applications as an alternative to petroleum-based plastic.

Bioplastic in Medical and Pharmaceuticals

Because of its various properties, bioplastic is applied in different medical aspects such as pharmacology, surgery, transplantology, drug delivery, and tissue engineering. Cells were grown *in-vitro* on biopolymer to create "tissue" for implantation in tissue engineering. It should be highly biocompatible as a general requirement for any foreign material incorporation into the human body. Silicone is a traditionally used polymer in the medical field. Still, silicon has some virulent effects on human health as it helps cancer cells to grow. In this case, biopolymers can play a role of a substitute for silicone [48]. Still, it has to pass through some physical requirements like the material needs to be biodegradable, biocompatible, and support cell adhesion and growth for tissue engineering. It should allow the development of cells, the passage of nutrients and waste products, and guide and organize the cells [49]. Some common bioplastics used in such applications are PLA, cellulose, and chitin-based. The bioplastic material should be smaller and diametrical on the nanoscale with high biodegradability for drug delivery. Also, biopolymer threads are commonly used in stitching wounds as they have biodegradable properties [50].

Bioplastic in Packaging Industries

Biodegradable materials are getting a positive approach as raw materials for packaging in different sectors. One of the most considerable polymer-producing industries is the packaging industry, where the food sector has the most significant marginal gaps [51]. During a study in 2016, the packaging industry was considered the most considerable application ground, covering around 40% of the total biopolymer market. It can be used in different sectors such as gift shops, food packaging, rapping, and any other packaging alternative for petroleum-based plastics. And due to its biodegradable property, it will not end up in landfills or oceans and stay there for hundreds of years. By this, there will be no clogged sewers and drains, polluting streets, beaches, and scenery, thus impacting waste management and its expenses [52].

Bioplastic in Domestic

Our domestic surroundings contain a large number of plastic households. At different periods, these things end up in landfills, or some products get burned, producing dangerous gases harmful for human health and the environment. A study in 2016 also concluded that there has been a significant growth in consumer and household goods, furniture, spore, *etc.*, which consume the 2nd most prominent portion of biopolymers [53].

Bioplastic in the Industry

Presently petroleum-based different plastic occupies the overall space in the industrial requirements. From paper clips to aeroplane manufacturing, plastic has been extensively used, which ends up in landfills and oceans. With people being aware of the disadvantages of plastic, different industries are shifting to better alternatives. For example, the usage of plastic straws has been depleted drastically in the last five years. Similarly, other industries are moving their manufacturing materials from petroleum-based plastic to bioplastic, such as cup industries, toothbrush, electrical, and fashion industries [54].

With time, other biopolymer applications will shine as fossil fuel usage leads to the shortage of plastic. Then the alternative bioplastic industry is going to resolve the requirements.

Biodegradation of Bioplastic in the Environment

With the increasing production of conventional plastics, over 250 million tons and its substitutes are of particularly microplastics. There has been an augmented pile-up of waste in terrestrial and marine environments. Bioplastic, an alternative, has revolutionized the worldwide market due to micro-organisms' sustainable properties and biodegradation ability. However, bioplastic is produced from sustainable renewable resources, and not all undergo biodegradation by micro-organisms. For instance, only polylactic acid-based and polyhydroxyalkanoates-based bioplastic are degraded by microbes. These microbes, upon degradation, assimilate polyesters of bioplastic and derive nutrients by biotic hydrolysis leaving behind byproducts like carbon dioxide, water molecules, and methane in the environment [55, 56]. Bioplastic degradation occurs with three processes: photo-degradation (light-based degradation), thermo-oxidative degradation, and biodegradation [57].

We talk about biodegradation and how micro-organisms can degrade biopolymers in their respective environment. Two factors, bioplastic kind and environment, often play a crucial role in biodegradation. However, other factors such as pH, temperature, time, and availability of nutrients, the intensity of UV radiation, oxygen concentration, and microbial diversity also play crucial roles in biodegradation processes [58, 59].

The process of biodegradation of bioplastic occurs in three significant steps (Fig. 4):

Stage of Biodeterioration

Bioplastic material undergoes fragmentation by micro-organisms to produce structured materials in this stage. It is characterized by various chemical, mechanical, and physical changes leading to cracks and fissures in bioplastics. The visible changes seen are weak discolouration and polymer splitting. High porosity influences the stage of biodeterioration [59, 60].

Stage of Biofragmentation

Micro-organism depolymerizes complex bioplastic into monomers and oligomers. The enzymes used for biodegradation belong to the proteinase, esterase, and cutinase families [59, 61]. The stage is also referred to as the stage of depolymerization. The enzymes are intra- and extracellular, depending on the micro-organisms and their habitat.

Stage of Assimilation

After fragmentation or depolymerization, micro-organisms assimilate or engulf monomers or oligomers to further degrade these to more specific products like organic and inorganic biomass. These later provide carbon, energy, and electron source for the growth of micro-organisms. Byproducts like carbon dioxide, methane, and water molecules are generated during the assimilation of monomers and oligomers.

Over ninety kinds of micro-organisms were identified as potentially biodegradable in bioplastics. These include aerobes, anaerobes, photosynthetic bacteria, archaebacteria, a lower class of eukaryotes, and fungi—these diversified micro-organisms have a habitat in soil, compost, and marine environment. The biodegradation process can occur in aerobic and anaerobic forms depending upon the nature and the habitat of the micro-organisms. During aerobic biodegradation, high oxygen (more than 6%) favours micro-organisms to use polymer as carbon, electron, and energy sources to produce end byproducts like carbon dioxide, water, and compost residue. However, during the anaerobic biodegradation, micro-organisms convert a polymer into methane, hydrogen sulfide, carbon dioxide, and hydrogen gases. The remaining is the digestate residue in an oxygen-free environment like a biogas digestor.

So, the biodegradation of bioplastic occurs depending on the polymer types, environmental factors, and diverse microbial groups [6]. Different microbial communities degrade various constituents forming bioplastics as well as the microbial communities present in the bioplastic waste also determine the biodegradation rate.

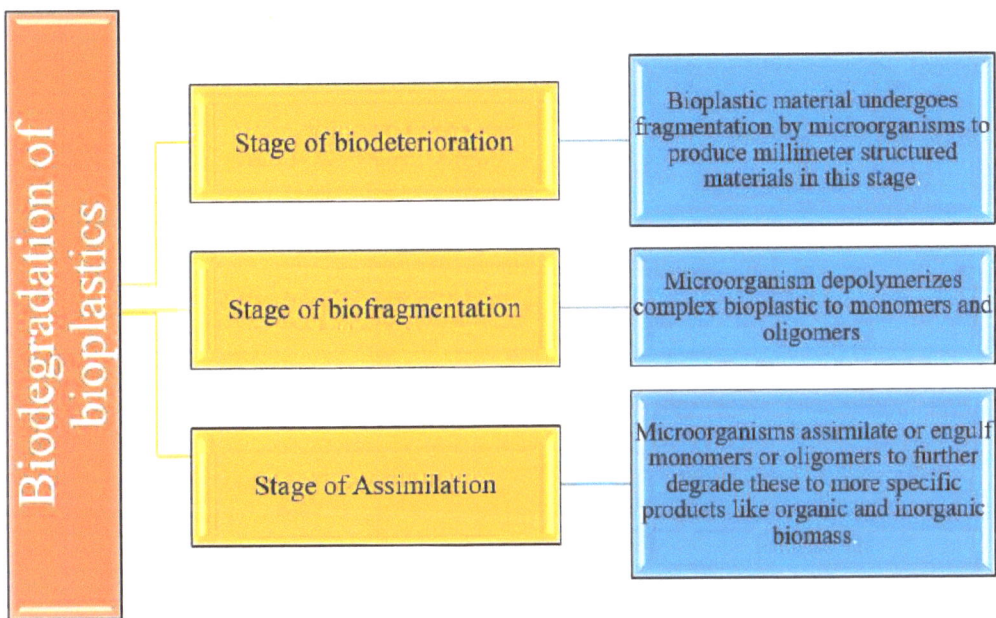

Fig. (4). Biodegradation of Bioplastics occurs in three stages.

CONCLUSION

The frequent microplastic contamination due to small particle size has shifted the paradigm towards bioplastic. Such microplastic contamination occurs due to the fragmentation of oversized plastics. Bioplastic, as an alternative, deals with contamination and overuse of petroleum or non-renewable resources in the production of plastics. There are plenty of renewable resources available for bioplastic production. However, microbial-based production has significantly increased in the past few decades. More and more research to enhance production is in progress. Of course, plant-based bioplastics are available in the market, but landfills are still a significant problem since they blend with synthetic plastic to provide durability. It is a large gap area that needs to be further explored for the complete biodegradation by micro-organisms. Micro-based bioplastics offer more advantages, such as being biocompatible, biodegradable to non-toxic byproducts, durable, and budgetary.

ACKNOWLEDGEMENT

Declared none.

REFERENCES

[1]   Narancic T, Verstichel S, Reddy Chaganti S, *et al.* Biodegradable plastic blends create new possibilities for end-of-life management of plastics but they are not a panacea for plastic pollution. Environ Sci Technol 2018; 52(18): 10441-52.
        [http://dx.doi.org/10.1021/acs.est.8b02963] [PMID: 30156110]

[2]   Andrady AL, Neal MA. Applications and societal benefits of plastics. Philos Trans R Soc Lond B Biol Sci 2009; 364(1526): 1977-84.
        [http://dx.doi.org/10.1098/rstb.2008.0304] [PMID: 19528050]

[3]   Yousuf A, Naseer M, Naqash N, Singh R. Isolation and identification of microplastic particles from agricultural soil and its detection by fluorescence microscope technique. Think India Journal 2019; 22(16): 3934-49.

[4]   Narancic T, O'Connor KE. Microbial biotechnology addressing the plastic waste disaster. Microb Biotechnol 2017; 10(5): 1232-5.
        [http://dx.doi.org/10.1111/1751-7915.12775] [PMID: 28714254]

[5]   Naqash N, Prakash S, Kapoor D, Singh R. Interaction of freshwater microplastics with biota and heavy metals: a review. Environ Chem Lett 2020; 18(6): 1813-24.
        [http://dx.doi.org/10.1007/s10311-020-01044-3]

[6]   Manzoor S, Naqash N, Rashid G, Singh R. Plastic material degradation and formation of microplastic in the environment: A review. Mater Today Proc 2021.

[7]   Roscam Abbing M. Plastic soup an atlas of ocean pollution. Plastic marine debris. Washington DC: Island Press 2019; p. 124.

[8]   Campanale C, Massarelli C, Savino I, Locaputo V, Uricchio VF. A detailed review study on potential effects of microplastics and additives of concern on human health. Int J Environ Res Public Health 2020; 17(4): 1212.
        [http://dx.doi.org/10.3390/ijerph17041212] [PMID: 32069998]

[9]   Jambeck J R, Geyer R, Wilcox C, *et al.* Plastic Waste Inputs from Land into the Ocean. Science 2015; 347(6223): 768-71.
        [http://dx.doi.org/10.1126/science.1260352]

[10]   Lebreton L C M, Van Der Zwet J, Damsteeg J W, Slat B, Andrady A, Reisser J. River plastic emissions to the world's oceans. Nat Commun 2017; 7: 15611.
        [http://dx.doi.org/10.1038/ncomms15611]

[11]   Meeker JD, Sathyanarayana S, Swan SH. Phthalates and other additives in plastics: human exposure and associated health outcomes. Philos Trans R Soc Lond B Biol Sci 2009; 364(1526): 2097-113.
        [http://dx.doi.org/10.1098/rstb.2008.0268] [PMID: 19528058]

[12]   Laskar N, Kumar U. Plastics and microplastics: A threat to environment. Environmental Technology & Innovation 2019; 14: 100352.
        [http://dx.doi.org/10.1016/j.eti.2019.100352]

[13]   Campanale C, Stock F, Massarelli C, *et al.* Microplastics and their possible sources: The example of Ofanto river in southeast Italy. Environ Pollut 2020; 258: 113284.

[http://dx.doi.org/10.1016/j.envpol.2019.113284] [PMID: 32005487]

[14] Smith M, Love DC, Rochman CM, Neff RA. Microplastics in seafood and the implications for human health. Curr Environ Health Rep 2018; 5(3): 375-86.
[http://dx.doi.org/10.1007/s40572-018-0206-z] [PMID: 30116998]

[15] Naqash N, Manzoor S, Singh R. Microplastic hazard, management, remediation, and control strategies: a review. Int J Environ Technol Manag 2022; 1(1): 1.

[16] Thiele C J, Hudson M D, Russell A E, Saluveer M, Sidaoui-Haddad G. Microplastics in fish and fishmeal: an emerging environmental challenge? Sci Rep 2021; 11(1): 2045.
[http://dx.doi.org/10.1038/s41598-021-81499-8]

[17] Nandiyanto ABD, Fiandini M, Ragadhita R, Sukmafitri A, Salam H, Triawan F. Mechanical and biodegradation properties of cornstarch-based bioplastic material. Materials physics and mechanics 2020; 44(3): 175.
[http://dx.doi.org/10.18720/MPM.4432020_9]

[18] Gill M. Bioplastic: a better alternative to plastics. Int J Res Appl Nat Soc Sci 2014; 2: 115-20.

[19] Reddy R L, Reddy V S, Gupta G A. Study of bio-plastics as green and sustainable alternative to plastics. Int J Emerging Technol Adv Eng 2013; 3(5): 76-81.

[20] Jan H. Bioplastics 2020; 1(4): 35.

[21] Ismail NA, Mohd Tahir S, Norihan Y, *et al.* Synthesis and characterization of biodegradable starch-based bioplastics. Mater Sci Forum 2016; 846: 673-8.
[http://dx.doi.org/10.4028/www.scientific.net/MSF.846.673]

[22] Bilo F, Pandini S, Sartore L, *et al.* A sustainable bioplastic obtained from rice straw. J Clean Prod 2018; 200: 357-68.
[http://dx.doi.org/10.1016/j.jclepro.2018.07.252]

[23] Arikan EB, Ozsoy HD. A review: investigation of bioplastics. Journal of Civil Engineering and Architecture 2015; 9(2).
[http://dx.doi.org/10.17265/1934-7359/2015.02.007]

[24] Bohmert-Tatarev K, McAvoy S, Daughtry S, Peoples OP, Snell KD. High levels of bioplastic are produced in fertile transplastomic tobacco plants engineered with a synthetic operon for the production of polyhydroxybutyrate. Plant Physiol 2011; 155(4): 1690-708.
[http://dx.doi.org/10.1104/pp.110.169581] [PMID: 21325565]

[25] Meereboer KW, Misra M, Mohanty AK. Review of recent advances in the biodegradability of polyhydroxyalkanoate (PHA) bioplastics and their composites. Green Chem 2020; 22(17): 5519-58.
[http://dx.doi.org/10.1039/D0GC01647K]

[26] Mostafa YS, Alrumman SA, Alamri SA, Otaif KA, Mostafa MS, Alfaify AM. Bioplastic (poly-3 hydroxybutyrate) production by the marine bacterium Pseudodonghicola xiamenensis through date syrup valorization and structural assessment of the biopolymer. Sci Rep 2020; 10(1): 8815-5.
[http://dx.doi.org/10.1038/s41598-020-65858-5] [PMID: 32483188]

[27] Dubey S, Bharmoria P, Gehlot PS, Agrawal V, Kumar A, Mishra S. 1-Ethyl-3-methylimidazolium diethylphosphate based extraction of bioplastic "polyhydroxyalkanoates" from bacteria: green and sustainable approach. ACS Sustain Chem& Eng 2018; 6(1): 766-73.
[http://dx.doi.org/10.1021/acssuschemeng.7b03096]

[28] Mudenur C, Mondal K, Singh U, Katiyar V. Production of polyhydroxyalkanoates and its potential applications.Advances in Sustainable Polymers. Springer 2019; pp. 131-64.
[http://dx.doi.org/10.1007/978-981-32-9804-0_7]

[29] Pantazaki AA, Tambaka MG, Langlois V, Guerin P, Kyriakidis DA. Polyhydroxyalkanoate (PHA) biosynthesis in Thermus thermophilus: purification and biochemical properties of PHA synthase. Mol Cell Biochem 2003; 254(1/2): 173-83.

[http://dx.doi.org/10.1023/A:1027373100955] [PMID: 14674696]

[30]   Benesova P, Kucera D, Marova I, Obruca S. Chicken feather hydrolysate as an inexpensive complex nitrogen source for PHA production by *Cupriavidus necator* on waste frying oils. Lett Appl Microbiol 2017; 65(2): 182-8.
[http://dx.doi.org/10.1111/lam.12762] [PMID: 28585326]

[31]   Abdo SM, Ali GH. Analysis of polyhydroxybutrate and bioplastic production from microalgae. Bull Natl Res Cent 2019; 43(1): 97.
[http://dx.doi.org/10.1186/s42269-019-0135-5]

[32]   Bustamante D, Segarra S, Tortajada M, *et al. In silico* prospection of microorganisms to produce polyhydroxyalkanoate from whey: *Caulobacter segnis*DSM 29236 as a suitable industrial strain. Microb Biotechnol 2019; 12(3): 487-501.
[http://dx.doi.org/10.1111/1751-7915.13371] [PMID: 30702206]

[33]   Bhatia SK, Kim JH, Kim MS, *et al.* Production of (3-hydroxybutyrate-co-3-hydroxyhexanoate) copolymer from coffee waste oil using engineered Ralstonia eutropha. Bioprocess Biosyst Eng 2018; 41(2): 229-35.
[http://dx.doi.org/10.1007/s00449-017-1861-4] [PMID: 29124334]

[34]   García G, Sosa-Hernández JE, Rodas-Zuluaga LI, Castillo-Zacarías C, Iqbal H, Parra-Saldívar R. Accumulation of PHA in the Microalgae *Scenedesmus* sp. under Nutrient-Deficient Conditions. Polymers (Basel) 2020; 13(1): 131.
[http://dx.doi.org/10.3390/polym13010131] [PMID: 33396913]

[35]   Muhammad Shamsuddin I, Ahmad Jafar J, Sadiq Abdulrahman Shawai A, Yusuf S, Lateefah M, Aminu I. Bioplastics as better alternative to petroplastics and their role in national sustainability: a review. Adv Biosci Bioeng (N Y) 2017; 5(4): 63.
[http://dx.doi.org/10.11648/j.abb.20170504.13]

[36]   Kaur H, Singh R. Analysis of nylon 6 as microplastic in harike wetland by comparing its ir spectra with virgin nylon 6 and 6.6. Eur J Mol Clin Med 2020; 7(07): 2020.

[37]   Ivleva NP, Wiesheu AC, Niessner R. Microplastic in aquatic ecosystems. Angew Chem Int Ed 2017; 56(7): 1720-39.
[http://dx.doi.org/10.1002/anie.201606957] [PMID: 27618688]

[38]   Issac MN, Kandasubramanian B. Effect of microplastics in water and aquatic systems. Environ Sci Pollut Res Int 2021; 28(16): 19544-62.
[http://dx.doi.org/10.1007/s11356-021-13184-2] [PMID: 33655475]

[39]   Gazal AA, Gheewala SH. Plastics, microplastics and other polymer materials-a threat to the environment. By J Sustain Energy Environ J Sustain Energy Environ 2020; 11: 113-22.

[40]   Manzoor S, Kaur H, Singh R. Existence of microplastic as pollutant in harike wetland: an analysis of plastic composition and first report on ramsar wetland of india. Curr World Environ 2021; 16(1).
[http://dx.doi.org/10.12944/CWE.16.1.12]

[41]   Mozaffari N, Kholdebarin A, Mozaffari NA. Review: investigation of plastics effect on the environment, bioplastic global market share and its future perspectives. Technog Ecol Saf 2019; 2019(5): 47-54.
[http://dx.doi.org/10.5281/ZENODO.2600664]

[42]   Narancic T, Cerrone F, Beagan N, O'Connor KE. Recent advances in bioplastics: application and biodegradation. Polymers (Basel) 2020; 12(4): 920.
[http://dx.doi.org/10.3390/polym12040920] [PMID: 32326661]

[43]   Büks F, Kaupenjohann M. Global concentrations of microplastics in soils – a review. Soil (Gottingen) 2020; 6(2): 649-62.
[http://dx.doi.org/10.5194/soil-6-649-2020]

[44]   Oluwasina OO, Akinyele BP, Olusegun SJ, Oluwasina OO, Mohallem NDS. Evaluation of the effects

of additives on the properties of starch-based bioplastic film. SN Applied Sciences 2021; 3(4): 421.
[http://dx.doi.org/10.1007/s42452-021-04433-7]

[45]   Wright SL, Thompson RC, Galloway TS. The physical impacts of microplastics on marine organisms:
        A review. Environ Pollut 2013; 178: 483-92.
        [http://dx.doi.org/10.1016/j.envpol.2013.02.031] [PMID: 23545014]

[46]   Witholt B, Kessler B. Perspectives of medium chain length poly(hydroxyalkanoates), a versatile set of
        bacterial bioplastics. Curr Opin Biotechnol 1999; 10(3): 279-85.
        [http://dx.doi.org/10.1016/S0958-1669(99)80049-4] [PMID: 10361079]

[47]   Choi J, Lee SY. Factors affecting the economics of polyhydroxyalkanoate production by bacterial
        fermentation. Appl Microbiol Biotechnol 1999; 51(1): 13-21.
        [http://dx.doi.org/10.1007/s002530051357]

[48]   Ulery BD, Nair LS, Laurencin CT. Biomedical applications of biodegradable polymers. J Polym Sci,
        B, Polym Phys 2011; 49(12): 832-64.
        [http://dx.doi.org/10.1002/polb.22259] [PMID: 21769165]

[49]   Verlinden RAJ, Hill DJ, Kenward MA, Williams CD, Radecka I. Bacterial synthesis of biodegradable
        polyhydroxyalkanoates. J Appl Microbiol 2007; 102(6): 1437-49.
        [http://dx.doi.org/10.1111/j.1365-2672.2007.03335.x] [PMID: 17578408]

[50]   Kehail AA, Brigham CJ. Anti-biofilm activity of solvent-cast and electrospun polyhydroxyalkanoate
        membranes treated with lysozyme. J Polym Environ 2018; 26(1): 66-72.
        [http://dx.doi.org/10.1007/s10924-016-0921-1]

[51]   Savitha R VYM, Savitha R. Overview on polyhydroxyalkanoates: a promising biopol. J Microb
        Biochem Technol 2011; 3(5): 99-105.
        [http://dx.doi.org/10.4172/1948-5948.1000059]

[52]   Bugnicourt E, Cinelli P, Lazzeri A, Alvarez V. Polyhydroxyalkanoate (PHA): Review of synthesis,
        characteristics, processing and potential applications in packaging. Express Polym Lett 2014; 8(11):
        791-808.
        [http://dx.doi.org/10.3144/expresspolymlett.2014.82]

[53]   Anjum A, Zuber M, Zia KM, Noreen A, Anjum MN, Tabasum S. Microbial production of
        polyhydroxyalkanoates (PHAs) and its copolymers: A review of recent advancements. Int J Biol
        Macromol 2016; 89: 161-74.
        [http://dx.doi.org/10.1016/j.ijbiomac.2016.04.069] [PMID: 27126172]

[54]   Cioica N, Cota C, Nagy M, Fodorean G. Plastics made from renewable sources – potential and
        perspectives for the environment and agriculture of the third millenium. Bull Univ Agric Sci Vet Med
        Cluj-Napoca Agric 1970; 65(2): 1843-5386.
        [http://dx.doi.org/10.15835/buasvmcn-agr:1083]

[55]   Thomas J, Thomas J. A methodological outlook on bioplastics from renewable resources. Open
        Journal of Polymer Chemistry 2020; 10(2): 21-47.
        [http://dx.doi.org/10.4236/ojpchem.2020.102002]

[56]   Atiwesh G, Mikhael A, Parrish CC, Banoub J, Le TAT. Environmental impact of bioplastic use: A
        review. Heliyon 2021; 7(9): e07918.
        [http://dx.doi.org/10.1016/j.heliyon.2021.e07918] [PMID: 34522811]

[57]   Bátori V, Åkesson D, Zamani A, Taherzadeh MJ, Sárvári Horváth I. Anaerobic degradation of
        bioplastics: A review. Waste Manag 2018; 80: 406-13.
        [http://dx.doi.org/10.1016/j.wasman.2018.09.040] [PMID: 30455023]

[58]   Emadian SM, Onay TT, Demirel B. Biodegradation of bioplastics in natural environments. Waste
        Manag 2017; 59: 526-36.
        [http://dx.doi.org/10.1016/j.wasman.2016.10.006] [PMID: 27742230]

[59]   Folino A, Karageorgiou A, Calabrò P S, Komilis D. Biodegradation of wasted bioplastics in natural

and industrial environments: a review. Sustainability 2020; 12(15): 6030.
[http://dx.doi.org/10.3390/su12156030]

[60]   Ruggero F, Belardi S, Caretti E, Lotti T, Lubello C, Gori R. Rigid and film bioplastics degradation under suboptimal composting conditions: a kinetic study. Waste Manag Res 2021.
[http://dx.doi.org/10.1177/0734242X211063731] [PMID: 34865591]

[61]   Sharma B, Jain P. Deciphering the advances in bioaugmentation of plastic wastes. J Clean Prod 2020; 275: 123241.
[http://dx.doi.org/10.1016/j.jclepro.2020.123241]

CHAPTER 9

Challenges to the Analysis of Microplastic Pollution from the Environment

Nafiaah Naqash[1] and **Rahul Singh**[1,*]

[1] *School of Bioengineering and Biosciences, Lovely Professional University, Phagwara-144411, Punjab, India*

Abstract: A growing interest in microplastic pollution in the environment demands simple, inexpensive, comparable, and robust methods for microplastic (MP) analysis. A wide range of methodologies for sampling, sample preparation, and MP analysis are in use. This chapter discusses the most common detection methods, as well as sampling strategies and sample preparation methods along with a special emphasis on challenges. The spectroscopic methods require time-consuming sample preparation and measurement durations, whereas thermo-analytical methods are faster but lack the ability to determine sample size distribution. Many articles concerning the quality and quantity of MPs in various matrices have been published. However, drawbacks and limitations in MP analyses are frequently overlooked or ignored. As a result, depending on the defined analytical question, the majority of the described methods are applicable. As a result, this chapter summarizes current sampling, sample preparation, and analysis methods, discusses limitations, and outlines the complexities associated with MP loss or contamination during sampling and laboratory testing.

Keywords: Characterization, Digestion, Detection, Density separation, Extraction, FTIR, Identification, Infrared spectroscopy, Limitations, Microplastic, Occurrence, Processing, Purification, Pyr-GC-MS, Quality, Quantification, Raman, SEM, Spectrometry, Sampling.

INTRODUCTION

Microplastics are ubiquitously found in diverse environmental systems, including marine waters, sediments, freshwater, and terrestrial ecosystems, including biota. Microplastics are a threat to the environment, and their presence, especially in water, has a negative influence on ecology and human health. They are discharged into the environment through a variety of sources, including ordinary plastic products, plastic degradation, industries, and wastewater treatment plants [1].

---
[*] **Corresponding author Rahul Singh:** School of Bioengineering and Biosciences, Lovely Professional University, Phagwara-144411, Punjab, India Email: rahulsingh.mlkzoology@gmail.com

Once they reach water, aquatic biota feeds on these toxins, and microplastic enters the food chain, posing serious health risks.

The analysis of microplastics from the environment is more challenging than larger plastic particles. Microplastics are pervasive, slow-degrading pollutants with characteristics such as durability, high stability, high fragmentation potential, and the ability to adsorb additional contaminants [2, 3]. Organic pollutants such as heavy metals, polyamide, polyester, polymerizing vinyl chloride, and acrylics have been found adhered to microplastics in previous investigations. Furthermore, it releases toxic compounds that are detrimental to humans as a result of trophic transfer from marine, freshwater, and terrestrial organisms. Although it is widely acknowledged that microplastics pose a threat to the environment, the numerous consequences of microplastics to the ecosystem have not been thoroughly investigated. In that case, it is critically important to enhance research techniques for analyzing microplastics from the environment. Microplastics should be analyzed as an emerging worldwide contaminant in order to assess potential impacts on the environment and humans [4].

Microplastics are a major challenge on their own, however, the first challenge to eliminate them involves the complete identification of their type, form, and morphology [4]. Plastic does not have a single molecular structure but is made up of a number of polymeric components of various sizes, shapes, and compositions with diverse additives. Further, from the methodological point of view, visual identification is inexpensive and relatively simple. However, this provides only a small portion of the total picture of microplastic abundance in the environment. More advanced techniques are necessary and were developed in recent decades to lower the limit of detection for microplastics in terms of particle size, false identification of particles as plastics, and, specifically to process larger sampling amounts in less duration for monitoring. Previous studies have considered bulk sample volume, sieving, filter pore size, density separation, and organic digestion for microplastic sampling. However, the use of novel methods, such as the enhancement of visual identification by staining dyes and the generalized use of chemical characterization, will improve sampling procedures for microplastics. Currently, the majority of research is focused on identifying and quantifying techniques of microplastics. The procedures include rapid screening techniques based on visual identification to a complex mix of analytical techniques that provide information on polymer type, particle number or mass, and particle size, which are extremely expensive and time-consuming. Though, the lack of standardized methods and protocols among researchers, and many approaches now in use may underestimate or overstate microplastic contamination, ensuing in contradictory data [5]. Therefore, it is mandatory to create standardized methodologies, such as sampling and identification methods for microplastics, to collect comparable monitoring data. In this chapter, we focused on the challenges experienced during primary methodologies and strategies for sampling, separating, identifying, and quantifying microplastics in the environment. The analytical procedures involved can provide basic descriptions of pollution levels, shifting patterns in microplastic concentrations, and the risk of organism exposure. To ensure the uniformity of microplastic data in the environment, the effectiveness and limitations of primary procedures and techniques for sampling, separating, detecting, and analyzing microplastics in the environment must be adopted as soon as possible.

ENVIRONMENTAL SAMPLING AND CHALLENGES

Microplastic has already been detected in freshwater, benthic sediment, soil, atmospheres, seas, and beach sand, as well as in far-flung locations such as the arctic regions and the Tibet Plateau. Microplastic can also be taken up by a variety of aquatic and terrestrial flora and fauna. As a result, microplastics may have a wide range of effects on the earth's ecosystem, and thus need to be addressed. Analyzing microplastics in the environment can offer fundamental information on pollution levels, changing patterns in microplastic concentrations, and the risk of organism exposure. The initial method of identifying and measuring microplastic contamination is to gather microplastic samples from major collecting areas including water, sediment, soil, and biota. The sampling procedures are obviously different, which ultimately alters the microplastic concentration.

Marine and Freshwater Sampling

To access microplastic contamination of marine and freshwater systems, large quantities of water are filtered through nets. Both freshwater and marine systems have similar sampling procedures, therefore, allowing for standardization of sampling methods in the future. However, different densities of 1.00 g/cm^3 and 1.03 g/cm^3 of freshwater and marine waters, respectively, could contribute to varying distributions of microplastics in the water column of each system [6]. Trawling is the most frequent type of water sampling, wherein Neuston and Manta trawl nets have been utilized most commonly. Furthermore, to estimate the sample volume, nets should be fitted with flow meters which allow the results to be expressed in m^3. Microplastic pollution in marine water across the Tropical Eastern Pacific and Galapagos was sampled using plankton nets with a 60 cm diameter, 3 m length, and 150 μm and 500 μm pore size. To avoid any oil or litter contamination from the main vessel, both nets were deployed simultaneously at a distance of 30 meters from the ship's stern [7]. Although, the Manta net allows sampling of large quantities of water; however, Plankton nets have also been employed due to their small mesh size (about 100 μm) with the capability of recovering 30 times more microplastics than manta trawl nets [8].

These techniques enable huge sample volumes by simultaneously taking into account a wider size range of MPs, and might be appropriate at various locations under various circumstances (*e.g.* lake, wastewater). Instead of nets, water pumps have been utilized for water intakes by vessel and deck pumps both in the case of coastal and freshwater sampling with different mesh sizes. Here, predetermined water amounts are either pumped *via* stacked steel sieves or connected filter cartridges [9]. Isokinetic sampling ensures a non-selective and homogeneous sampling of particles by withdrawing water at the same rate as the water flow. Additionally, size classes down to 10 m can be separated locally using cascade-based filtration. Another alternative is a continuous flow centrifuge that retains all particles with densities greater than 1 g/cm^3. However, it must be investigated to see if using this method, MPs can be quantitatively sampled.

Pumps with attached filter cascades are less mobile than nets. Additionally, the risk of fragmentation, particle loss (*e.g.* hose system) and contamination (*e.g.* abrasion of components) during sampling has to be acknowledged and validated. However, the comparison of different methods clearly shows that quantitative and representative sampling with a net is only possible for samples of larger MPs. Another difficulty occurs with the comparability of results. Sampled volumes differ and so do extrapolated MP concentrations (MPs per liter, m^3, mass units or particle numbers). Moreover, reporting sampling-related limitations (*e.g.* size range) and appropriate units is mandatory for comparable data. In the future, a combined sampling approach might be probably the most practical solution to sample a wide size range of MPs quantitatively. For instance, MPs >300 μm are sampled by nets, and simultaneously the size fraction <300 μm is sampled by a pump-filter combination.

Sediment and Soil Sampling

Several extraction methods are applied for identification and characterization of MPs from soil and sediments such as, manual sorting [10], oil extraction [11], and density separation [12]. The extraction methods are followed by purification for the removal of organic components by hydrogen peroxide [13], Fenton's reagent, and/or enzymatic digestion [14]. The use of strong acidic solutions during the purification alters plastic composition by degrading the labile plastics. The identification methods also vary considerably from visual identification, staining with fluorescent Nile red, GC-MS, pyrolysis GC-MS, and thermal extraction desorption (TED) GC-MS [15]. There is a high possibility of losing the particle size, number, and shape related information while using these methods. Focal plane array-based μ-FTIR, and Raman μ-spectroscopy are useful in identifying the polymer type, shape, size, and number within any environmental sample [16].

Sampling from soil/sediment is done with grabbers and corers for attaining the deep layer samples from the soil profile. Grabbers allow sampling from the upper layer whereas, core sampling allows sampling from the undisturbed cores [17]. Thus, in-depth MP deposition and age analysis is potentially preserved. The sample number is limited in undisturbed sampling achieved with cores [18]. There are higher chances of sediment loss during drilling to more depths. Bioturbation also reduces the validity of MP-accumulation dependent on age [19]. Grab sampling ensures large volume sample retrieval but disturbs sediments [20]. Besides, the limitations associated with the collecting device, the approaches also deter the comparison in relation to sampling depth or mass or reference units. There are no standardized protocols for easy and effective MP-sample retrieval, there is urgent need for standardization of MP-sampling protocols governing soil/sediment for the generation of comparable and representative data [20, 21]. During MP sampling, it is also imperative to avoid contact of samples with any plastic equipment. Since there are several intrinsic challenges in collected MP samples from the environment directly, MPs used in laboratory are often acquired commercially. This subjects the test organism to uniform plastic exposure rather than what is present in the environment. More often than not, the anti-aggregation agents are used that affect the physiochemical parameters of plastic.

Air and Biota Sampling

Air is a critical carrier for MP transportation and can reach up to a distance of 95 kms [22]. In a study, indoor and outdoor air was investigated with 2-355 particles/m^2 atmospheric fallout [23]. In another study, more than 350 MPs per m^2 were reported in French pyrenees [22]. Vianello and his colleagues examined human exposure to indoor MPs and reported 2-16 MPs per m^3 [24]. Mostly MP samples from air are taken using a pump at 1.2 m height for several hours [25]. However, pump-based sampler systems with a uniform flow rate are expensive to maintain. These are largely effected by the environmental fluctuations. An alternative sampler system used are the moistened filters [26]. These are cheap and easy to maintain but they require long time for sampling. Besides, the indoor air samples contain abundant polymer-based particles that outweigh the particlesof inorganic origin. There is a need to establish plastic free laboratories for precise and accurate MP analysis [27].

MPs are ubiquitous, and interactions with a wide variety of species that live in land and aquatic settings are inevitable. So, theoretically, a survey of the biota may include a huge variety of species. The appropriate compartment and biocoenosis determine the sample techniques used. For instance, grabbers and nets are frequently used to sample benthic and pelagic organisms, respectively. The size of the organisms determines the sampling strategy and scale. The majority of studies have concentrated on a variety of aquatic species, including fish, mussels, and planktonic invertebrates [28, 29, 30]. Despite these concerns about ingesting microplastic, little is known about the consequences on the environment and how they might affect the food chain. Because there aren't any adequate standardised methodologies, data are frequently unrepresentative, incomparable, and lacking in quality control. The lack of evidence concerning MPs' intake in larger organisms is a result of expensive, time-consuming techniques and ethical ambiguity. Therefore, MPs are monitored in stranded carcasses or faeces of larger species including birds, cetaceans, and seals [31, 32, 33]. Large-scale sampling, particularly of invertebrates, will increase the representativeness of the results. This suggests that the investigation of MP contamination in biota should only focus on species with high local densities. It is necessary to identify temporal and spatial differences when considering the dynamic architecture of biocenoses. To avoid MP egestion during transport and storage, on-site fixation of the collected specimens with preservative chemicals (for example, ethanol) is advised. Specimens should be checked for attached MPs before sample preparation (such as digestion of biota). Overall, data on MP abundances in abiotic compartments are supplemented by sampling of the biota. In addition to existing monitoring efforts, representative, easily accessible important species should be picked and studied.

Only a small portion of the species have been sampled and examined for MP contamination, despite the richness of the biotic environment [34]. For instance, nothing is known about how much nano-plastics (NP) have contaminated bacteria, algae, macrophytes, terrestrial plants, and vertebrates in particular. Results from biota reflect only chosen (*e.g.*, size, shape) and time-dependent (*e.g.*, egestion) MP abundances in the environment since MP ingestion depends on both the bioavailability of MPs and the feeding type and size selectivity of biota. It is difficult to link environmental MP contaminations directly to ingested microplastics in terms of quantities. Identification of these restrictions might be made easier by characterizing MPs' intake through laboratory experiments.

Due to excessive stress, the use of fixatives may cause abrupt evacuation of the gastrointestinal tract. Prior to sampling, this needs to be taken into account and tested. The examination of combined samples from different species is generally hampered by species-specific potentials to ingest MPs. Comparing laboratory and environmental results, however, is challenging because MP concentrations in the field are frequently some orders of magnitude higher than those observed in laboratory research with biota. As a result, we still lack complete information about the effects and fate of microplastics. There is an urgent need for a standardized protocol for conducting studies on the ingestion of microplastics by biota to address this issue because the area of microplastic research is still in its infancy and work done now provides the groundwork for future research [35].

SAMPLE PREPARATION AND CHALLENGES

Sample pretreatment is an important step in detecting MP contamination. Where, macro-plastics can be easily detected in the environment, MP detection requires extensive processing of native samples. The complexity of the matrix varies between and within the environmental compartments. For example, the concentration and composition of suspended solids depend on the body of water and location (groundwater, lakes, streams, oceans, *etc.*). The same applies to the composition of sediments (such as beaches and lakeshores), soils (rich in organic matter or minerals), biota samples (such as biomass, proteins, fats, carbohydrate content, biological structure), and air. The above variations imply different sample preparation requirements, as particulate matter (minerals, organics, *etc.*) interferes with the identification of MPs.

With increasing proportions of particulate matrix matter, the complexity of sample preparation increases. However, reducing the volume of the native sample is the first step toward representative counts. This is accomplished by distinguishing MPs from natural components based on material properties. Several methods have been proposed, including sieving, density and oil separation, electro separation, and organic material digestion [36]. For sample preparation, single or multiple combined techniques (*e.g.*, sieving, density separation, and organic material digestion) are used depending on the investigated matrices. Native sample processing frequently entails altering the original conditions. Sample preparation, for example, can change MP sizes, polymer composition, hetero-aggregates, biofilms, MP degradation state, and sorbed pollutants [37]. Furthermore, the variety of methods limits the comparability of studies. The following section contains detailed information on each method:

Size and Density Separation

Size fractionation *via* sieving, which is used on aqueous and sediment samples, may allow for easier processing of environmental samples in the subsequent extraction and purification steps. Microscopy and spectroscopic analysis of MPs are limited by variable and wide particle sizes. For example, different particle sizes necessitate different magnifications, and large particles may obscure small particles. To fractionate MP size classes for an optimised analysis using microscopy and spectroscopy, stacked sieves with different mesh sizes can be used [37]. Wet or dry sieving is possible here. Closed wet sieving units eliminate dust formation (*e.g.*, particle loss) and contamination. Sieve clogging can cause particle loss and inaccurate fractionation. Furthermore, abrasion or fragmentation during sieving may change the number, size, and shape of MPs. Small particles are washed away during wet sieving with a small mesh size if the washing fluid is not filtered to recover small fractions.

Plastics are lightweight materials that can be distinguished from natural components by using densities differences. This situation is commonly used to separate MPs from sediment or soil matrices. Commodity plastic polymers, for example, range in density from 0.01 to >1.4 g/cm^3 (*e.g.* foamed PS and PVC), whereas mineral materials are typically denser (*e.g.* >2.5 g/cm^3 for quartz, feldspar, and calcite) [38]. MPs float to the top and minerals sink to the bottom when environmental matrices are suspended in high-density liquids (>1.2 g/cm^3). The spectrum of recoverable polymers is determined by the density of the separation fluid. Furthermore, monitoring the density of separation fluids is critical for reproducibility.

Density separation is accomplished using a variety of devices, ranging from simple conical flasks and funnels to custom-built machinery (*e.g* [39].). The Munich Plastic Sediment Separator, for example, can process large sample volumes in a single run (max. 6 l sediment) [39]. Recovery rates of 95.5 percent and 100 percent for small and large MPs, respectively, are additional advantages, as is the closed construction. Following density separation, the supernatant is filtered (pressure or vacuum) and/or centrifuged before being processed further. Large MPs in the filtered supernatant can be picked up with tweezers and sorted under a stereomicroscope. Fluidization is another possibility. Nuelle *et al.* [13] reduced the mass of a sediment sample by introducing turbulent gas bubbles generated by air into a density solution and then adding the sediment. Lighter particles rose to the surface and were transferred to another glass vessel.

Although differences in densities are a promising and widely used method for separating plastic materials from natural components, several limitations must be recognized. For starters, the densities of organic matter are comparable to those of plastic materials. As a result, organic matter and other materials with similar densities will float in dense liquids and be included in the sample. Second, the density of plastics is affected by MP degradation and the formation of biofilms and hetero-aggregates. Furthermore, particle aggregation and disaggregation during density separation may influence the fate of MPs (*e.g.* sedimentation of MPs attached to larger minerals). Third, method validation is mostly done with synthetic sediments. As a result, recovery rates do not apply directly to complex environmental matrices. Fourth, the variety of methods makes it difficult to compare results (*e.g.* equipment used, sample volume and separation liquid).

Alternative Extraction Methods

MPs can also be separated from other materials using centrifugal forces. A paper on sink-float density separation of polyolefins from waste was published by Bauer *et al.* [40]. For the first separation of larger waste, a two-stage processing with a centrifugal force separator and a hydro jig was used. Because of the oleophilic properties of plastics, oil extraction is another option for separating MPs from environmental samples. Crichton *et al.* [41] swirled a dry sample with water and canola oil several times to ensure that each particle was in contact with oil. The sample settled, and the oil was decanted, rinsed, and filtered before being incubated in reagent alcohol to remove any remaining oil on the MPs. The recovery rate for expanded polystyrene (EPS), polyvinyl chloride (PVC), acrylonitrile butadiene styrene (ABS), vinyl, polyamide (PA) fibres, and polyester fibres was greater than 90%. (PES). This method is inexpensive and simple to use. Plastics are nonconductive materials, whereas sediments and particulate matter are conductive [42]. As a result, electroseparation can be used to separate sediments and MPs. The method is based on particle electrostatic behaviour [43]. Dry particles are separated into conductive and nonconductive particles after being charged with a corona electrode up to 30 kV. The recovery rate of microplastic particles >63 m in beach sediments which is close to 100 percent. This method allows for the processing of large sample volumes while reducing the remaining matrix to 10%. Because of the material reduction, only trace amounts of chemicals are required. Furthermore, it is a relatively inexpensive method because only the device is necessary and no chemicals are required.

Centrifugation, oil separation, and electroseparation have yet to be tested and validated for the separation of a wide range of polymers, size classes, and shapes. Furthermore, additional research is required to validate recovery rates for various matrices and environmental MPs (*e.g.* hetero-aggregates). For centrifugation, one can expect expensive apparatuses and high maintenance costs because the rotating parts may be subjected to significant force and sensitivity to the efficiency of the approach when applied to samples from various environments. High concentrations of organic matter in oil separation may reduce the overall efficiency. Separating sediment components based on conductivity is only effective in completely dry samples. As a result, particle loss, particularly in the smaller-size fractions, is possible and must be considered in mitigation strategies.

Purification

Certain acids and bases have a high resistance to synthetic polymers. This is beneficial for MP purification. In theory, incubating environmental samples with acids and/or bases causes organic matter digestion, resulting in volume reduction. However, whether or not MPs are unaffected is dependent on the chemicals used and the type of polymer used. Polyesters like PET, PBT, PC, PLA, as well as other synthetic polymers like cellulose acetate (CA) and PVC, can be degraded with 10 M NaOH [44, 45]. The use of HCl results in high digestion efficiency (>95%), albeit at the expense of melting synthetic polymers such as PET [46]. Digestion with H_2O_2 and Fenton's reagent has been widely used, with encouraging results. However, a study found that after a week of incubation with 30% H_2O_2, only 70% of MPs were recovered [47]. Another gentle method for removing organic materials without affecting synthetic polymers is enzymatic digestion with single or sequential enzyme incubation [48]. Depending on the study and validated workflows, samples are either treated before or after density separation. In general, digestion duration and efficiency are influenced by the complexity of the sample (water, sediment, biota, or air).

Synthetic polymers are damaged by strong acids and alkaline solutions. The degree of degradation and discoloration is polymer-specific and related to the chemicals used. Acid digestion with chemicals such as HCl and HNO_3 is not suggested because synthetic polymers such as nylon and PET are sensitive to these acids. Plastic fragments may also be effected or discolored by alkaline digestion [6]. Moreover, certain compromises must be made in order to effectively remove organic matter. This includes chemical concentration, incubation temperature, and time. Even though enzymatic digestion does not dissolve or degrade synthetic polymers and produces excellent purification results, the protocols proposed are time-consuming [44, 48]. To effectively digest biological structures, for example, several sequentially applied enzymes (*e.g.*, lipase, chitinase, lignase, and proteinase) are required (*e.g.* lipids, chitin carapaces, lignin, carbohydrates and proteins). Furthermore, exposure conditions must be monitored and adjusted for optimal reaction rates.

Visual Identification

In previous studies, many researchers used visual identification with microscopes to identify MP particles. Visual, light, or digital microscopes are commonly used for analysing larger plastic particles ranging in size from 300 μm to 5 mm (*e.g* [49, 50]). Smaller particles are not as easily identified as plastics, so they are examined under a stereomicroscope first, and synthetic polymers are later identified using Raman, pyrolysis GC-MS, or FTIR spectroscopy [51, 52]. Sorting chambers, such as Bogorov counting chambers, can be used to sort aqueous samples. Size limits for visual inspection without the assistance of other characterization techniques are recommended to be greater than 500 µm [53] or even greater than 1 mm. Norén [54] suggests a standard criterion for visual identification of particles to reduce the possibility of misidentification, such as the absence of organic origin structures on MPs, equal thickness in MP fibres, homogeneous coloration of the particle, and finally the use of fluorescence microscopy to exclude biological origin structures.

Larger particles can be identified as possible plastic fragments based on their shape, size, degradation stage, and colour [38], but there is no guarantee of identification. Smaller particles, particularly those smaller than 300 m, are difficult to identify because they may be mineral or organic in origin [38, 55]. Due to their small size, particles smaller than a certain size can become unmanageable when handled with instruments such as forceps . Furthermore, the results of visual identification are subjective, dependent on the sample matrix, and time-consuming. As a result, visual identification should be used only as a preliminary evaluation of results, in conjunction with another method for greater accuracy and precision [56]. The approach should not be used on MP particles closer than 500 m because the possibility of misidentification is very high. Rather it is highly urged to use spectroscopic approaches to aid in the accurate identification of such MPs [57].

INSTRUMENTAL ANALYSIS OF MICROPLASTIC

Particle analysis and characterization are classified into two types. Sizes (distribution), shapes, and colours are considered in morphological or physical classification, whereas polymer types and additives are considered in chemical classification. The morphological and chemical composition of plastic particles are frequently studied using spectroscopy and mass spectrometry. Another technique for studying particle morphology is scanning electron microscopy (SEM). SEM, on the other hand, cannot distinguish polymer particles from natural particles based on spectral information. A detailed overview of the various approaches used in MP analysis, from sampling and sample treatment to different detection methods, has been provided.

Fourier Transform Infrared (FTIR) and Raman Spectroscopy

For decades, FTIR spectroscopy has been used for the analysis and characterization of synthetic organic polymers and their products [58]. FTIR spectroscopy is an absorption technique in which the IR radiation absorption by molecule vibration is determined by the change in the dipole moment of a chemical bond within the sample molecule. The sample is irradiated with infrared light (mid-IR range), and a portion of the IR radiation is absorbed by the exited molecule vibrations within the sample being probed and detected either by reflection or transmission mode. Because of their ever-repeating molecule composition, synthetic polymers have highly specific IR spectra with distinct signals, making it an ideal technique for MP identification. They can be identified by comparing their spectra to reference spectra and characterised based on their chemical structure. The surface technique attenuated total reflectance (ATR) FTIR spectroscopy allows for the fast and reliable single analysis of large MP particles >500 μm [58, 59], but reliable results require a relatively clean surface, and potential MP particles with biofilm must be cleaned by wiping with alcohol, for example.

Micro-FTIR spectroscopy, which combines FTIR spectroscopy with an IR microscope, allows for the analysis of particles as small as 10μm. Transmission mode measurements necessitate the use of IR transparent filters (*e.g.*, aluminium oxide or silica) on which the sample is placed. The total absorption of IR radiation through samples above a certain thickness limits the sample thickness that can be analysed with this mode. Micro-FTIR mapping with a single element detector and focal plane array (FPA)-based FTIR imaging have been widely used to identify MPs. The former takes a long time, whereas the latter allows for the acquisition of thousands of spectra in a short period of time and at a high spatial resolution.

This aids in the sequential imaging of the entire sample filters and is currently the most widely used method for identifying and characterising MPs. It is also possible to use reflectance mode in FTIR. However, because particles with irregular shapes cause refractive error, this mode produces complex spectra that are difficult to interpret [58, 59]. Overall, the advantage of FTIR analysis is that it provides qualitative and quantitative information about each MP particle while preserving the sample for further downstream analysis. This method evades extrapolation of measured data from sample filter subareas as well as potential uncertainties due to unequal particles size on the filter. However, assessing entire filters generates large datasets with up to one million IR spectra, which are best analysed using automated methods [60, 61].

This method does not provide information on the number of MPs. One of the major drawbacks of this technique is that due to the diffraction limit of light, very small particles as small as 10 μm cannot be analysed. It is especially difficult to analyse fibres with diameters in that range. Furthermore, there have been reports of significant underestimation of MP particles with a diameter of 20 μm [62]. Particle thicknesses greater than 50-100 μm result in total absorption as well as black particles that absorb strongly in the IR range. If such particles are present, data analysis may be difficult and MPs may be underestimated. Water is a strong absorber of IR radiation, so samples must be thoroughly dried. Other limitations are shared by the complementary FTIR and Raman spectroscopy techniques. These will be covered in the section following Raman spectroscopy.

Raman, like FTIR spectroscopy, provides a unique spectral fingerprint of different chemical structures; however, the main difference is that Raman spectroscopy is dependent on changes in the polarizability of the chemical bond within molecules constituting a sample. As a result, the two approaches complement each other; signals that are strong in IR may be weaker in Raman spectra, and *vice versa*. To interact with the sample, a monochromatic light source, such as a laser, is used. The wavelengths of the most commonly used lasers range from 500 to 1,064 nm. Radiation from this source interacts with the sample, and one out of every billion photons from the source is in-elastically scattered, revealing information about the sample's molecular vibrations [63]. The Raman shift is the difference in frequency between inelastically scattered photons and Rayleigh photons (photons that do not interact with the sample) that forms the basis of the Raman spectrum [64]. A synthetic polymer's Raman spectrum contains several distinct sharp signals that correspond to the chemical functional groups that make up the sample and can be identified by comparing it to reference spectra. At the moment, the micro-Raman spectroscopy approach (a Raman spectrometer setup coupled with a microscope) is widely used. A confocal mode of configuration ensures lateral and depth resolution, allowing the analysis of particles as small as 0.5-1 m depending on the type of sample being examined. In comparison to FTIR spectroscopy, a higher percentage of MPs can thus be detected in terms of size. Raman spectroscopic imaging mode allows for qualitative and quantitative estimation of small size-range MPs, though time duration is the primary constraint because Raman images are acquired through stepwise point measurements, known as mapping [64].

Raman spectroscopy can also be used to analyse and localise specific components such as MPs within complex matrices such as biological cells and tissues. Furthermore, it is an important method for monitoring changes in the biochemistry of organisms when stressors such as MPs are ingested. In recent years, micro-Raman spectroscopy has been used worldwide for the analysis of samples ranging from single microbial cells to MPs [65, 66]. Although this method allows for the detection of even the smallest MP particles in environmental samples, time-effective integration for MP research has yet to be demonstrated [52, 23]. Raman has a significant limitation in terms of fluorescence. Surface alterations ranging from humic substance sorption to surface oxidation (ageing) or biofouling are a major source of fluorescence in MP analysis. Numerous changes to the particle surface are likely to occur during the dwell time of MP particles in (aqueous) environments. Furthermore, artificial dyes in coloured plastic particles can obstruct detection and result in the generation of dye spectra rather than polymer spectra. More sample preparation steps, as described in the previous section, are required to avoid fluorescence problems [67]. Humic substances can also be removed by using bases such as potassium hydroxide and sodium hydroxide solutions.

Pyrolysis Gas Chromatography Mass Spectrometry (Pyr-GC-MS)

Pyrolysis GC-MS (Pyr-GC-MS) enables the quantification of MP masses in ambient matrices in contrast to FTIR and Raman spectroscopy. By adjusting the pyrolysis temperature, certain volatile degradation products that resemble or may be linked to their synthetic polymer precursors are produced [62]. By using gas chromatography (GC) column to separate these pyrolysis byproducts, mass spectrometry (MS) findings can be used to identify them. Pyrolysis fingerprints make it easier to characterise and identify the synthetic polymer that makes up MPs. These pyrograms can be matched to known, virgin synthetic polymer sample pyrograms used as references. Thermochemolysis, or thermally assisted methylation and hydrolysis, is used for synthetic polymers containing polar subunits like polyesters. This process is known to increase sensitivity, enhance chromatographic separation, and provide more structural details. If a thermal desorption step is utilised before pyrolysis, plastic additives can also be identified concurrently during Pyr-GC-MS analysis. Previously, MPs from environmental materials such as marine sediments, river sediments, sea water surface, and marine creatures have been analysed by Pyr-GC-MS [62, 68, 69]. The concentration of an entire environmental MP collection on filters after purification and the subsequent Pyr-GC-MS analysis were recently demonstrated by Fischer and Scholz-Böttcher [70].This method is promising because optical or mechanical presorting was not required [71]. Additionally, utilising calibration techniques, mass-related measurement was achieved at trace levels. Dierkes *et al.* [72] demonstrated that it was possible to measure the mass of the most popular polymer types, PE, PP, and PS, with limits of quantification as low as 0.007 mg/g sample.

Pyr-GC-MS has several restrictions when it comes to MP quantitation. First off, because samples are pyrolyzed, analysis is destructive. Second, because quantification is mass-based, it is impossible to determine the number or shape of the particles. Instead of a unique pyrolysis product for copolymers, analysis of copolymers would provide pyrolysis products of each co-monomer. Furthermore, although they are not particularly included in mass-based measurement, additives, fillers, and colours contribute to the mass of MP particles. In reality, it is quick and easy to identify individual MP particles using particular pyrolysis products. Mass-based assessment of MPs in environmental samples, and consequently, a wide range of MP particles, is a lot more difficult. In the literature [73, 74], specific pyrolysis products for popular synthetic polymers are extensively documented. However, the availability of organic materials such as hydrophobic chemicals (fats and waxes) and natural polymers (cellulose, keratin, chitin, and lignin) leads to similar or identical pyrolysis products, which may cause the concentration of MP to be overestimated.

Thermal Extraction and Desorption-Gas

Mass spectrometry and chromatography (TED-GC-MS) and the TED-GC-MS method combine thermal sample extraction with a thermogravimetric analysis (TGA) of solid-phase adsorbers, followed by thermal desorption-gas chromatography mass spectrometry (TDS-GC-MS) examination of the adsorbers [74, 75]. TED-GC-MS is the name given to this combo. This method involves heating the entire sample to up to 1,000 °C in the TGA before pyrolyzing it. On a solid-phase adsorber, the degradation products peculiar to synthetic polymers are adsorbed. These products must differ from the degradation products of the ambient matrix. The majority of synthetic polymers start to break down at 350 degrees Celsius, making it possible to remove ambient matrix elements that break down at much lower temperatures and are not stuck on a solid phase adsorber. These important plastic degradation byproducts are captured on a solid-phase adsorber like polydimethylsiloxane (PDMS). Thermal desorption is used to desorb the trapped decomposition products, which are then transferred by an inert gas, such as helium, into a cooled injection system. Additionally, only substances that can be thermally desorbed at temperatures between 200 and 300°C are evaporated for GC-MS analysis. This serves as a filter by obstructing the majority of contamination products unique to long-chain polymers from entering the GC-MS [76]. Identification and quantification of samples like MPs from various environments are made possible by analysis of these products by GC-MS. Based on the presence of mass fragments typical of various decomposition products unique to that polymer, the identity of a synthetic polymer is confirmed. Thus, MPs in environmental samples can be quickly identified and quantified using TED-GC-MS. But as of now, only a very small number of plastic kinds have been confirmed to be examined by TED-GC-MS in environmental samples (*e.g.* PE, PP, PS, PA 6 and PET [77]).

This method has yet to be thoroughly tested and implemented for the analysis of MPs from various environmental samples. The Pyr-GC-MS technique uses samples, however they are destroyed before they can be used for downstream analysis. Additionally, only 20 mg of dry sample may be analysed during a single TED-GC-MS run; as a result, multiple consecutive runs will be required to analyse a large number of samples. However, it is indicated that a sample containing up to 100 mg can be analysed in a matter of 2-3 hours. Due to the sample's total homogeneity, MP particle size, shape, colour, and associated information are irrelevant in this method. Initial research revealed a detection limit of roughly 0.5 to 1 weight percent in a 20 mg sample, which appears to be quite high. Additionally, calibration for a complete array of pertinent synthetic polymers has not yet been accomplished or demonstrated.

Scanning Electron Microscopy Coupled with Energy Dispersive X-Ray Spectroscopy (SEM-EDX)

SEM-EDX, a technique that combines scanning electron microscopy and energy dispersive X-ray spectroscopy, can be used to gather high-resolution data on the morphology and in-depth details on the chemical make-up of samples like MPs. The surface of the samples under investigation can be seen in high-magnification, high-resolution photographs thanks to SEM [77]. The application of different detectors by SEM microscopes produces images based on the emission and detection of secondary electrons. Additionally, based on atomic number, the backscattered electron detector provides data on topography and material contrast (Z). SEM can be connected to EDX, which uses element-specific radiation to characterise the surface near volume's chemical composition.

This method offers elemental analysis on regions with a diameter as tiny as a nanometer. The sample is exposed to an electron beam, which causes X-rays that are indicative of the elements present on the sample. The intensities provide quantitative data on the distribution and composition of the elements. As a result, the method may be used to gather comprehensive data on the elemental composition of MPs as well as data on the inorganic additives that are present in the sample. Furthermore, the presence of materials like biomass on MPs can be implicated in the strong signals from specific elements, such as nitrogen. In comparison to optical microscopy, SEM-EDX is a promising method for obtaining high-quality images of MPs and then determining their elemental composition. These details can be used to analyse nonplastic and synthetic polymer-based materials. For tracking MP ocalization in biota, it can be utilised as a supplemental approach to fluorescence microscopy.

Since EDX information is only based on elemental composition, SEM-EDX is not suited for differentiating between different plastic polymers in environmental samples. It is also a very expensive procedure that necessitates a great deal of knowledge, time, and time-consuming sample preparation. Additionally, conducting elements like Au, AuPd, or carbon must be sprayed onto nonconducting MP samples. Samples cannot, therefore, be used for downstream analysis. Working with low kV beam energies, such as 0.5–1.5 kV, could, however, eliminate the requirement for sputtering samples. As a result, sample preparation might go more quickly, and analysis might go much more smoothly.

Quantitative Nuclear Magnetic Resonance Spectroscopy (1 H-qNMR) and MALDI-TOF Mass Spectrometry

An easy, quick, and nondestructive method that enables the simultaneous characterisation and quantification of many molecules is quantitative nuclear magnetic resonance (1 H-qNMR). It has been widely applied in numerous research and development domains, including polymers, metabolites, medicines, forensic sciences, environmental applications, *etc* [78]. Access to a wealth of detailed knowledge about the structure and behaviour of complex molecules is made possible by 1 H-qNMR. This method investigates variations in the magnetic field that each atom's nucleus in a molecule experiences when subjected to an external magnetic field. 1 H and 13 C are two examples of odd-mass or odd-atomic-number nuclei that have nuclear spins, and the spins of their nuclei can be detected by NMR experiments. Modern equipment can accurately and effectively analyse even complicated molecules like proteins since they are sensitive to local magnetic fluctuations as small as one part in a billion, which can provide insight about the molecular structure [78]. Depending on the proportional relationship between the integrated signal area and the number of resonant nuclei, 1 H-qNMR provides information. There are numerous ways of determination that can be used, including calibration curve method, standard addition, absolute determination using internal and external benchmarks, and relative determination. The size-independent descriptive and analytical characterisation and identification of MP particles from various matrices can be accomplished using 1 H-qNMR. Since the precise concentration of all analytes apart from polymers in an aquatic environment should not be known, preliminary investigations employed for the analysis of MPs such as PE, PET, and PS have demonstrated that the calibration curve method is most suited for MP analysis [79]. As a result, 1 H-qNMR is an exact quantification technique that may be used to obtain a quantification accuracy of >98%. It is a quick and efficient strategy (about 1 min per sample measurement). Using the calibration curve approach, a high throughput analysis is possible.

Significant problems could develop as a result of environmental factors like bacteria and other bioorganic elements. These components' errors are signal overlays that cannot be fixed, and systematic errors are challenging to find.

Therefore, to eliminate all these components prior to analysis, a successful sample preparation step is required. The requirement to dissolve analytes in an appropriate deuterated solvent, which results in the loss of MP particle size information, is a significant disadvantage of this method. Due to the intrinsic chemical and physical characteristics of polymers in contrast to other organic compounds, it is crucial that a suitable solvent be found for the dissolving of various forms of synthetic polymers. The solvent signal may additionally overlap or coincide with the signal range of the relevant synthetic polymers, which can complicate the study greatly. Although 1 H-qNMR is a promising method for the size-independent study of MPs, there are still a number of parameters that need to be tuned in the context of MP analysis and are being used for this purpose only in the very early stages.

Dimzon *et al*. [80] and Weidner *et al*. [81] investigated the suitability of MALDI-TOF MS for detecting various polymers using various sample preparation techniques. A limitation is that different polymers require different cationizing agents for ionisation. The ability to detect different polymer materials with high selectivity makes use of MALDI-TOF MS in MP analysis appealing. Direct detection of additives or sorbed pollutants on particles is one possible application. Through imaging techniques, MALDI-TOF spectra can provide morphological information such as particle size and shape. Rivas *et al*. [82] investigated polymer surface modifications during degradation using MALDI-TOF mass spectrometry imaging (MSI). The matrix was sublimated and deposited into the target *via* a special coating chamber after samples were placed on indium-tin oxide glass slides. The spectra were recorded, and various m/z ranges of interest were chosen to generate MSI. This technique determines the spatial changes of a polymer surface.

CONCLUSION

Numerous challenges exist in sampling, identification, and quantification of microplastics in various environmental systems. There is a need to develop efficient and detailed sampling strategies, as sampling is critical for accurate results. Furthermore, sampling of these particles does not account for seasonal or inter-annual variations in environmental parameters. Spectroscopic techniques are still appropriate for large particles and will find use in particle analysis of micro- and macro-sized particles. As it is now for special experimental setups, additional methods such as SEM and fluorescence microscopy can be used. Moreover, standardised sampling and sample preparation protocols are still lacking, resulting in significant biases when only small sample portions are analysed. Analytical methods, such as sampling and sample preparation, require evidence, establishing comparable methodologies. These considerations may eventually lead to the implementation of standardised methodologies for sampling and quantifying micro- and nanoplastics in the environment. Only then will the collected data allow for a thorough assessment of these materials' potential ecotoxicological effects, actively contributing to the filling of these knowledge gaps.

ACKNOWLEDGEMENTS

Authors sincerely acknowledge and express their gratitude to the School of Bioengineering and Biosciences, Lovely Professional University, India.

REFERENCES

[1]   Vivekanand A, Mohapatra S, Chemosphere VT. Microplastics in aquatic environment: Challenges and perspectives.Chemosphere 2021; 282: 131151.
[http://dx.doi.org/10.1016/j.chemosphere.2021.131151]

[2]   Padervand M, Lichtfouse E, Robert D, Wang C. Removal of microplastics from the environment. A review. Environ Chem Lett 2020; 18(8).
[http://dx.doi.org/10.1007/s10311-020-00983-1]

[3]   Naqash N, Prakash S, Kapoor D, Singh R. Interaction of freshwater microplastics with biota and heavy metals: a review. Environ Chem Lett 2020; 18(6): 1813-24.
[http://dx.doi.org/10.1007/s10311-020-01044-3]

[4]   Manzoor S, Kaur H, Singh R. Existence of microplastic as pollutant in harike wetland: an analysis of plastic composition and first report on ramsar wetland of india article history Curr World Environ 2021; 16(1).
[http://dx.doi.org/10.12944/CWE.16.1.12]

[5]   Prata JC, da Costa JP, Duarte AC, Rocha-Santos T. Methods for sampling and detection of microplastics in water and sediment: A critical review. Trends Analyt Chem 2019; 110: 150-9.
[http://dx.doi.org/10.1016/j.trac.2018.10.029]

[6]   Alfaro-Núñez A, Astorga D, Cáceres-Farías L, *et al.* Microplastic pollution in seawater and marine organisms across the Tropical Eastern Pacific and Galápagos. Sci Rep 2021; 11(1): 6424.
[http://dx.doi.org/10.1038/s41598-021-85939-3]

[7]   Dris R, Gasperi J, Rocher V, Saad M, Renault N, Tassin B. Microplastic contamination in an urban area: a case study in Greater Paris. Environ Chem 2015; 12(5): 592-9.
[http://dx.doi.org/10.1071/EN14167]

[8]   Lenz R, Labrenz M. Small microplastic sampling in water: development of an encapsulated filtration device. Water 2018; 10(8): 1055.
[http://dx.doi.org/10.3390/w10081055]

[9]   Piehl S, Leibner A, Löder MGJ, Dris R, Bogner C, Laforsch C. Identification and quantification of macro- and microplastics on an agricultural farmland. Sci Rep 2018; 18(8): 17950.
[http://dx.doi.org/10.1038/s41598-018-36172-y]

[10]   Scopetani C, Chelazzi D, Mikola J, *et al.* Olive oil-based method for the extraction, quantification and identification of microplastics in soil and compost samples. Sci Total Environ 2020; 733: 139338.
[http://dx.doi.org/10.1016/j.scitotenv.2020.139338] [PMID: 32446078]

[11]   Möller JN, Löder MGJ, Laforsch C. Finding microplastics in soils: a review of analytical methods. Environ Sci Technol 2020; 54(4): 2078-90.
[http://dx.doi.org/10.1021/acs.est.9b04618] [PMID: 31999440]

[12]   Nuelle MT, Dekiff JH, Remy D, Fries E. A new analytical approach for monitoring microplastics in marine sediments. Environ Pollut 2014; 184: 161-9.

[http://dx.doi.org/10.1016/j.envpol.2013.07.027] [PMID: 24051349]

[13] Löder MGJ, Imhof HK, Ladehoff M, *et al.* Enzymatic purification of microplastics in environmental samples. Environ Sci Technol 2017; 51(24): 14283-92.
[http://dx.doi.org/10.1021/acs.est.7b03055] [PMID: 29110472]

[14] Schrank I, Möller JN, Imhof HK, *et al.* Microplastic sample purification methods - Assessing detrimental effects of purification procedures on specific plastic types. Sci Total Environ 2022; 833: 154824.
[http://dx.doi.org/10.1016/j.scitotenv.2022.154824] [PMID: 35351498]

[15] Klein S, Dimzon IK, Eubeler J, Knepper TP. Analysis, occurrence, and degradation of microplastics in the aqueous environment. Handbook of Environmental Chemistry 2018; 58: 51-67.
[http://dx.doi.org/10.1007/978-3-319-61615-5_3]

[16] Pojar I, Stănică A, Stock F, Kochleus C, Schultz M, Bradley C. Sedimentary microplastic concentrations from the Romanian Danube River to the Black Sea. Sci Rep 2021; 11(1): 2000.
[http://dx.doi.org/10.1038/s41598-021-81724-4]

[17] Tsuchiya M, Nomaki H, Kitahashi T, Nakajima R, Fujikura K. Sediment sampling with a core sampler equipped with aluminum tubes and an onboard processing protocol to avoid plastic contamination. MethodsX 2019; 6: 2662-8.
[http://dx.doi.org/10.1016/j.mex.2019.10.027] [PMID: 31799135]

[18] Waldschläger K, Brückner MZM, Carney Almroth B, *et al.* Learning from natural sediments to tackle microplastics challenges: A multidisciplinary perspective. Earth Sci Rev 2022; 228: 104021.
[http://dx.doi.org/10.1016/j.earscirev.2022.104021]

[19] Miller E, Sedlak M, Lin D, *et al.* Recommended best practices for collecting, analyzing, and reporting microplastics in environmental media: Lessons learned from comprehensive monitoring of San Francisco Bay. J Hazard Mater 2021; 409: 124770.
[http://dx.doi.org/10.1016/j.jhazmat.2020.124770] [PMID: 33450512]

[20] Yousuf A, Naseer M, Journal NN-TI. Isolation and identification of microplastic particles from agricultural soil and its detection by fluorescence microscope technique. Think india journal 2019; 22(16).

[21] Frias JPGL, Otero V, Sobral P. Evidence of microplastics in samples of zooplankton from Portuguese coastal waters. Mar Environ Res 2014; 95: 89-95.
[http://dx.doi.org/10.1016/j.marenvres.2014.01.001] [PMID: 24461782]

[22] Allen S, Allen D, Phoenix VR, *et al.* Atmospheric transport and deposition of microplastics in a remote mountain catchment. nature geoscience 2019.
[http://dx.doi.org/10.1038/s41561-019-0335-5]

[23] Dris R, Gasperi J, Saad M, Mirande C, Tassin B. Synthetic fibers in atmospheric fallout: A source of microplastics in the environment? Mar Pollut Bull 2016; 104(1-2): 290-3.
[http://dx.doi.org/10.1016/j.marpolbul.2016.01.006] [PMID: 26787549]

[24] Vianello A, Jensen RL, Liu L, Vollertsen J. Simulating human exposure to indoor airborne microplastics using a Breathing Thermal Manikin. Scientific Reports 2019; 9: 8670.
[http://dx.doi.org/10.1038/s41598-019-45054-w]

[25] Wright SL, Gouin T, Koelmans AA, Scheuermann L. Development of screening criteria for microplastic particles in air and atmospheric deposition: critical review and applicability towards assessing human exposure. Microplastics and Nanoplastics. Springer 2021.
[http://dx.doi.org/10.1186/s43591-021-00006-y]

[26] Harrold Z, Arienzo MM, Collins M, *et al.* A peristaltic pump and filter-based method for aqueous microplastic sampling and analysis. ACS ES&T Water 2022; 2(2): 268-77.
[http://dx.doi.org/10.1021/acsestwater.1c00270]

[27] Stanton T, Johnson M, Nathanail P, MacNaughtan W, Gomes RL. Freshwater and airborne textile

fibre populations are dominated by 'natural', not microplastic, fibres. Sci Total Environ 2019; 666: 377-89.
[http://dx.doi.org/10.1016/j.scitotenv.2019.02.278] [PMID: 30798244]

[28]   Prata JC, da Costa JP, Lopes I, Duarte AC, Rocha-Santos T. Effects of microplastics on microalgae populations: A critical review. Sci Total Environ 2019; 665: 400-5.
[http://dx.doi.org/10.1016/j.scitotenv.2019.02.132] [PMID: 30772570]

[29]   Foekema EM, De Gruijter C, Mergia MT, van Franeker JA, Murk AJ, Koelmans AA. Plastic in north sea fish. Environ Sci Technol 2013; 47(15): 8818-24.
[http://dx.doi.org/10.1021/es400931b] [PMID: 23777286]

[30]   Setälä O, Norkko J. Feeding type affects microplastic ingestion in a coastal invertebrate community. Mar Pollut Bull 2016; 102(1): 95-101.
[http://dx.doi.org/10.1016/j.marpolbul.2015.11.053]

[31]   Besseling E, Foekema EM, Van Franeker JA, *et al.* Microplastic in a macro filter feeder: Humpback whale Megaptera novaeangliae. Mar Pollut Bull 2015; 95(1): 248-52.
[http://dx.doi.org/10.1016/j.marpolbul.2015.04.007] [PMID: 25916197]

[32]   Nelms S, Galloway T, Godley B, *et al.* Investigating microplastic trophic transfer in marine top predators.Elsevier n.d. 2018.

[33]   Rebolledo E, Franeker J, van , *et al.* Plastic ingestion by harbour seals (phoca vitulina).LibraryWurNl n.d. 2013.

[34]   Collard F, Gasperi J, Gabrielsen GW, Tassin B. Plastic particle ingestion by wild freshwater fish: a critical review. Environ Sci Technol 2019; 53(22): 12974-88.
[http://dx.doi.org/10.1021/acs.est.9b03083] [PMID: 31664835]

[35]   Hermsen E, Mintenig SM, Besseling E, Koelmans AA. Quality criteria for the analysis of microplastic in biota samples: a critical review. Environ Sci Technol 2018; 52(18): 10230-40.
[http://dx.doi.org/10.1021/acs.est.8b01611] [PMID: 30137965]

[36]   Primpke S, Christiansen SH, Cowger W, *et al.* Critical assessment of analytical methods for the harmonized and cost-efficient analysis of microplastics. Appl Spectrosc 2020; 74(9): 1012-47.
[http://dx.doi.org/10.1177/0003702820921465] [PMID: 32249594]

[37]   Lusher AL, Munno K, Hermabessiere L, Carr S. Isolation and Extraction of Microplastics from Environmental Samples: An Evaluation of Practical Approaches and Recommendations for Further Harmonization. Appl Spectrosc 2020; 74(9): 1049-65.
[http://dx.doi.org/10.1177/0003702820938993] [PMID: 32893667]

[38]   Hidalgo-Ruz V, Gutow L, Thompson RC, Thiel M. Microplastics in the marine environment: a review of the methods used for identification and quantification. Environ Sci Technol 2012; 46(6): 3060-75.
[http://dx.doi.org/10.1021/es2031505] [PMID: 22321064]

[39]   Imhof HK, Schmid J, Niessner R, Ivleva NP, Laforsch C. A novel, highly efficient method for the separation and quantification of plastic particles in sediments of aquatic environments. Limnol Oceanogr Methods 2012; 10(7): 524-37.
[http://dx.doi.org/10.4319/lom.2012.10.524]

[40]   Bauer M, Lehner M, Schwabl D, *et al.* Sink–float density separation of post-consumer plastics for feedstock recycling. J Mater Cycles Waste Manag 2018; 20(3): 1781-91.
[http://dx.doi.org/10.1007/s10163-018-0748-z]

[41]   Crichton EM, Noël M, Gies EA, Ross PS. A novel, density-independent and FTIR-compatible approach for the rapid extraction of microplastics from aquatic sediments. Anal Methods 2017; 9(9): 1419-28.
[http://dx.doi.org/10.1039/C6AY02733D]

[42]   Köhnlechner R, Sander S. Praktischer Einsatz elektrostatischer Separatoren in der Sekundärrohstoffindustrie. BHM Berg- Und Hüttenmännische Monatshefte. Springer 2009; 154: pp.

136-9.
[http://dx.doi.org/10.1007/s00501-009-0453-2]

[43] Felsing S, Kochleus C, Buchinger S, Brennholt N, Stock F, Reifferscheid G. A new approach in separating microplastics from environmental samples based on their electrostatic behavior. Environ Pollut 2018; 234: 20-8.
[http://dx.doi.org/10.1016/j.envpol.2017.11.013] [PMID: 29154206]

[44] Hurley RR, Lusher AL, Olsen M, Nizzetto L. Validation of a method for extracting microplastics from complex, organic-rich, environmental matrices. Environ Sci Technol 2018; 52(13): 7409-17.
[http://dx.doi.org/10.1021/acs.est.8b01517] [PMID: 29886731]

[45] Dehaut A, Cassone AL, Frère L, *et al.* Microplastics in seafood: Benchmark protocol for their extraction and characterization. Environ Pollut 2016; 215: 223-33.
[http://dx.doi.org/10.1016/j.envpol.2016.05.018] [PMID: 27209243]

[46] Karami A, Golieskardi A, Choo CK, Romano N, Ho YB, Salamatinia B. A high-performance protocol for extraction of microplastics in fish. Sci Total Environ 2017; 578: 485-94.
[http://dx.doi.org/10.1016/j.scitotenv.2016.10.213] [PMID: 27836345]

[47] Lusher AL, McHugh M, Thompson RC. Occurrence of microplastics in the gastrointestinal tract of pelagic and demersal fish from the English Channel. Mar Pollut Bull 2013; 67(1-2): 94-9.
[http://dx.doi.org/10.1016/j.marpolbul.2012.11.028] [PMID: 23273934]

[48] Löder MGJ, Imhof HK, Ladehoff M, *et al.* Enzymatic purification of microplastics in environmental samples. Environ Sci Technol 2017; 51(24): 14283-92.
[http://dx.doi.org/10.1021/acs.est.7b03055] [PMID: 29110472]

[49] Eriksen M, Maximenko N, Thiel M, *et al.* Plastic pollution in the south pacific subtropical gyre. Mar Pollut Bull 2013; 68(1-2): 71-6.
[http://dx.doi.org/10.1016/j.marpolbul.2012.12.021] [PMID: 23324543]

[50] Lattin GL, Moore CJ, Zellers AF, Moore SL, Weisberg SB. A comparison of neustonic plastic and zooplankton at different depths near the southern California shore. Mar Pollut Bull 2004; 49(4): 291-4.
[http://dx.doi.org/10.1016/j.marpolbul.2004.01.020] [PMID: 15341821]

[51] Lenz R, Enders K, Stedmon CA, Mackenzie DMA, Nielsen TG. A critical assessment of visual identification of marine microplastic using Raman spectroscopy for analysis improvement. Mar Pollut Bull 2015; 100(1): 82-91.
[http://dx.doi.org/10.1016/j.marpolbul.2015.09.026] [PMID: 26455785]

[52] Cole M, Lindeque P, Fileman E, *et al.* Microplastic ingestion by Zooplankton... - Google Scholar n.d. Available at:https://scholar.google.com/scholar?&q=Cole (accessed on: 2022) 2013, 47, 12, 6646-6655.

[53] Catarino AI, Thompson R, Sanderson W, Henry TB. Development and optimization of a standard method for extraction of microplastics in mussels by enzyme digestion of soft tissues. Environ Toxicol Chem 2017; 36(4): 947-51.
[http://dx.doi.org/10.1002/etc.3608] [PMID: 27583696]

[54] Sweeden KS-NRR by K. 2007 undefined. Small plastic particles in Coastal Swedish waters. Researchgate Net n.d.

[55] Lavers J, Oppel S. Factors influencing the detection of beach plastic debris.Elsevier n.d. 2016.

[56] Vianello A, Boldrin A, Guerriero P, *et al.* Microplastic particles in sediments of Lagoon of Venice, Italy: First observations on occurrence, spatial patterns and identification. Estuar Coast Shelf Sci 2013; 130: 54-61.
[http://dx.doi.org/10.1016/j.ecss.2013.03.022]

[57] Dekiff J, Remy D, Klasmeier J. Occurrence and spatial distribution of microplastics in sediments from NorderneyElsevier n.d. 2014.

[58]   Löder MGJ, Gerdts G. Methodology used for the detection and identification of microplastics—a critical appraisal. Mar Anthropog Litter 2015; 201-27.
[http://dx.doi.org/10.1007/978-3-319-16510-3_8]

[59]   Manzoor S, Singh R, Kaur H. Clinical medicine analysis of nylon 6 as microplastic in harike wetland by comparing its ir spectra with virgin nylon 6 and 6.6 article in. Eur J Transl Clin Med 2021.

[60]   Harrison JP, Ojeda JJ, Romero-González ME. The applicability of reflectance micro-Fourie--transform infrared spectroscopy for the detection of synthetic microplastics in marine sediments. Sci Total Environ 2012; 416: 455-63.
[http://dx.doi.org/10.1016/j.scitotenv.2011.11.078] [PMID: 22221871]

[61]   Primpke S, Lorenz C, Rascher-Friesenhausen R, Gerdts G. An automated approach for microplastics analysis using focal plane array (FPA) FTIR microscopy and image analysis. Anal Methods 2017; 9(9): 1499-511.
[http://dx.doi.org/10.1039/C6AY02476A]

[62]   Hufnagl B, Steiner D, Renner E, Löder MGJ, Laforsch C, Lohninger H. A methodology for the fast identification and monitoring of microplastics in environmental samples using random decision forest classifiers. Anal Methods 2019; 11(17): 2277-85.
[http://dx.doi.org/10.1039/C9AY00252A]

[63]   Käppler A, Fischer M, Scholz-Böttcher BM, *et al.* Comparison of μ-ATR-FTIR spectroscopy and py-GCMS as identification tools for microplastic particles and fibers isolated from river sediments. Anal Bioanal Chem 2018; 410(21): 5313-27.
[http://dx.doi.org/10.1007/s00216-018-1185-5] [PMID: 29909455]

[64]   Applications of Vibrational Spectroscopy in the Study of Explosives n.d. Available at:https://apps.dtic.mil/sti/citations/ADA408892(accessed July 26, 2022).

[65]   Kuhar N, Sil S, Verma T, Umapathy S. Challenges in application of Raman spectroscopy to biology and materials. RSC Advances 2018; 8(46): 25888-908.
[http://dx.doi.org/10.1039/C8RA04491K] [PMID: 35541973]

[66]   Stöckel S, Kirchhoff J, Neugebauer U, Rösch P, Popp J. The application of Raman spectroscopy for the detection and identification of microorganisms. J Raman Spectrosc 2016; 47(1): 89-109.
[http://dx.doi.org/10.1002/jrs.4844]

[67]   Araujo CF, Nolasco MM, Ribeiro AMP, Ribeiro-Claro PJA. Identification of microplastics using Raman spectroscopy: Latest developments and future prospects. Water Res 2018; 142: 426-40.
[http://dx.doi.org/10.1016/j.watres.2018.05.060] [PMID: 29909221]

[68]   Djaker N, Lenne PF, Marguet D, Colonna A, Hadjur C, Rigneault H. Coherent anti-Stokes Raman scattering microscopy (CARS): Instrumentation and applications. Nucl Instrum Methods Phys Res A 2007; 571(1-2): 177-81.
[http://dx.doi.org/10.1016/j.nima.2006.10.057]

[69]   Fries E, Dekiff JH, Willmeyer J, Nuelle MT, Ebert M, Remy D. Identification of polymer types and additives in marine microplastic particles using pyrolysis-GC/MS and scanning electron microscopy. Environ Sci Process Impacts 2013; 15(10): 1949-56.
[http://dx.doi.org/10.1039/c3em00214d] [PMID: 24056666]

[70]   Hermabessiere L, Himber C, Boricaud B, *et al.* Optimization, performance, and application of a pyrolysis-GC/MS method for the identification of microplastics. Anal Bioanal Chem 2018; 410(25): 6663-76.
[http://dx.doi.org/10.1007/s00216-018-1279-0] [PMID: 30051208]

[71]   Fischer M, Scholz-Böttcher BM. Simultaneous trace identification and quantification of common types of microplastics in environmental samples by pyrolysis-gas chromatography–mass spectrometry. Environ Sci Technol 2017; 51(9): 5052-60.
[http://dx.doi.org/10.1021/acs.est.6b06362] [PMID: 28391690]

[72]   Ter Halle A, Jeanneau L, Martignac M, *et al.* Nanoplastic in the north atlantic subtropical gyre. Environ Sci Technol 2017; 51(23): 13689-97.
[http://dx.doi.org/10.1021/acs.est.7b03667] [PMID: 29161030]

[73]   Dierkes G, Lauschke T, Becher S, Schumacher H, Földi C, Ternes T. Quantification of microplastics in environmental samples *via* pressurized liquid extraction and pyrolysis-gas chromatography. Anal Bioanal Chem 2019; 411(26): 6959-68.
[http://dx.doi.org/10.1007/s00216-019-02066-9] [PMID: 31471683]

[74]   David J, Weissmannová HD, Steinmetz Z, *et al.* Introducing a soil universal model method (SUMM) and its application for qualitative and quantitative determination of poly(ethylene), poly(styrene), poly(vinyl chloride) and poly(ethylene terephthalate) microplastics in a model soil. Chemosphere 2019; 225: 810-9.
[http://dx.doi.org/10.1016/j.chemosphere.2019.03.078] [PMID: 30904761]

[75]   Dümichen E, Eisentraut P, Bannick CG, Barthel AK, Senz R, Braun U. Fast identification of microplastics in complex environmental samples by a thermal degradation method. Chemosphere 2017; 174: 572-84.
[http://dx.doi.org/10.1016/j.chemosphere.2017.02.010] [PMID: 28193590]

[76]   Dümichen E, Barthel AK, Braun U, *et al.* Analysis of polyethylene microplastics in environmental samples, using a thermal decomposition method. Water Res 2015; 85: 451-7.
[http://dx.doi.org/10.1016/j.watres.2015.09.002] [PMID: 26376022]

[77]   Duemichen E, Braun U, Senz R, Fabian G, Sturm H. Assessment of a new method for the analysis of decomposition gases of polymers by a combining thermogravimetric solid-phase extraction and thermal desorption gas chromatography mass spectrometry. J Chromatogr A 2014; 1354: 117-28.
[http://dx.doi.org/10.1016/j.chroma.2014.05.057] [PMID: 24929909]

[78]   Gniadek M, Dąbrowska A. The marine nano- and microplastics characterisation by SEM-EDX: The potential of the method in comparison with various physical and chemical approaches. Mar Pollut Bull 2019; 148: 210-6.
[http://dx.doi.org/10.1016/j.marpolbul.2019.07.067] [PMID: 31437623]

[79]   Peez N, Becker J, Ehlers SM, *et al.* Quantitative analysis of PET microplastics in environmental model samples using quantitative ^{1}H-NMR spectroscopy: validation of an optimized and consistent sample clean-up method. Anal Bioanal Chem 2019; 411(28): 7409-18.
[http://dx.doi.org/10.1007/s00216-019-02089-2] [PMID: 31489440]

[80]   Dimzon IKD, Knepper TP. MALDI–TOF MS for characterization of synthetic polymers in aqueous environment. Compr Anal Chem 2012; 58: 307-38.
[http://dx.doi.org/10.1016/B978-0-444-53810-9.00008-0]

[81]   Weidner SM, Trimpin S. Mass spectrometry of synthetic polymers. Anal Chem 2010; 82(12): 4811-29.
[http://dx.doi.org/10.1021/ac101080n] [PMID: 20491451]

[82]   Rivas D, Ginebreda A, Pérez S, Quero C, Barceló D. MALDI-TOF MS Imaging evidences spatial differences in the degradation of solid polycaprolactone diol in water under aerobic and denitrifying conditions. Sci Total Environ 2016; 566-567: 27-33.
[http://dx.doi.org/10.1016/j.scitotenv.2016.05.090] [PMID: 27213667]

CHAPTER 10

Distribution of Microplastics in Man-made Water Bodies

Suraya Partap Singh[1,*] and **Reetika Rani**[2]

[1] *Department of Zoology, Government Degree College, Boys, Kathua, Jammu & Kashmir, India*

[2] *Department of Zoology, School of Bioengineering and Biosciences, Lovely Professional University, Phagwara, Punjab, India*

Abstract: Plastic is one of the most commonly produced and used materials in the world due to its great features. It has also become the most prevalent type of debris found in our oceans, lakes, wetlands, and other lentic systems. Plastic (from the Greek "plastikos", meaning mouldable) is made of synthetic organic polymers. Anthropogenic activity has resulted in the deposition of a complex combination of materials in different water bodies, which may include synthetic polymers (plastics) which are degraded into smaller fragments which will be in the size of <5 mm; these are termed microplastics. Microplastic pollution is one of the main matters of concern nowadays, specifically due to the increasing anthropogenic activities in and around the different water bodies which lead to ubiquitous distribution of microplastics in water systems. It is a gleaming topic among the environmentalists of the world. The environmental release of MPs will occur from a wide variety of sources, including emissions from wastewater treatment plants, cosmetics, toothpaste, *etc.* and from the degradation of larger plastic debris. In recent years, interest in the effects of microplastics (MPs) has shifted towards freshwater ecosystems and in this chapter, we provide an overview of the issues of microplastic pollution that are concerned with manmade water bodies which can be inland as well as coastal environments as well as the sources of contamination of water bodies with microplastics, their influence and a conclusion.

Keywords: Aquatic, Coastal, Contamination, Cosmetics, Debris, Degradation, Environment, Items, Inland, Lake, Lentic, Manmade, Microplastics, Monomers, Ocean, Plastic, Pollution, Treatment, Water bodies, Wastewater.

INTRODUCTION

Waterbodies are defined as systems that have an accumulation of water on the surface of the planet Earth. The study of freshwater inland waters including lakes ponds, rivers, springs, streams, and wetlands is termed Limnology, and Oceano-

---

* **Corresponding author Suraya Partap Singh:** Department of Zoology, Government Degree College, Boys, Kathua, Jammu & Kashmir, India; E-mail: suraya9@gmail.com

Rahul Singh and Neeta Raj Sharma (Eds.)

graphy deals with the study of ocean water, it covers a wide range of topics, including marine life and ecosystems, ocean circulation, plate tectonics and the geology of the seafloor, and the chemical and physical properties of the ocean. Waterbodies are known by a plethora of different names in English: rivers, streams, ponds, bays, gulfs, seas, *etc.*

Generally, water bodies are naturally occurring bodies but with the modifications of living standards and increasing mental capability, human beings started building artificial water bodies for certain purposes. Such water bodies which are maintained by anthropogenic activities are termed Man-made water bodies. It started with the building of small ponds in villages for certain purposes like passing the dry spell to small tanks for storing water and reaching up to huge reservoirs which can be large-sized dams, reservoirs, large-sized tanks, abandoned quarries, *etc.* There are a number of artificial water bodies. The US accounts for approximately 20% of the standing water area under artificial water bodies and their impact on hydrology, sedimentology, geochemistry, and ecology is apparently large in proportion to their area [1]. In India, there are maximum inland water bodies as compared to the coastal type of water bodies. According to the Ministry of Jal Shakti of India, the Inland water resources of the country are classified as rivers and canals, and reservoirs, which include tanks and ponds. The brackish water lakes are also present in India. Except rivers and canals, total water bodies cover about 7 million hectares of the geographical area of India. Even the country experiences serious water scarcity problems due to the unequal spatial distribution of water resources and high population density. To cope up with this, India has constructed a number of wetlands which are shown in Fig. (1) where the wastewater is treated for reuse. The source and quality of the wastewater are also important aspects that are being observed properly while maintaining artificial wetlands in developing countries. The waste-water containing industrial wastes, domestic and organic wastes is largely used for constructing such wetlands [2]. The main aim of constructing these wetlands is to do advanced treatment of waste-water so that better quality water can be used for different purposes. The ecosystem of the artificial water bodies brings froth various advantages to residents of that area which are enlisted as a source of nutrition, hydration, and clean water for the use of human beings even for irrigation and other agricultural perspectives [3].

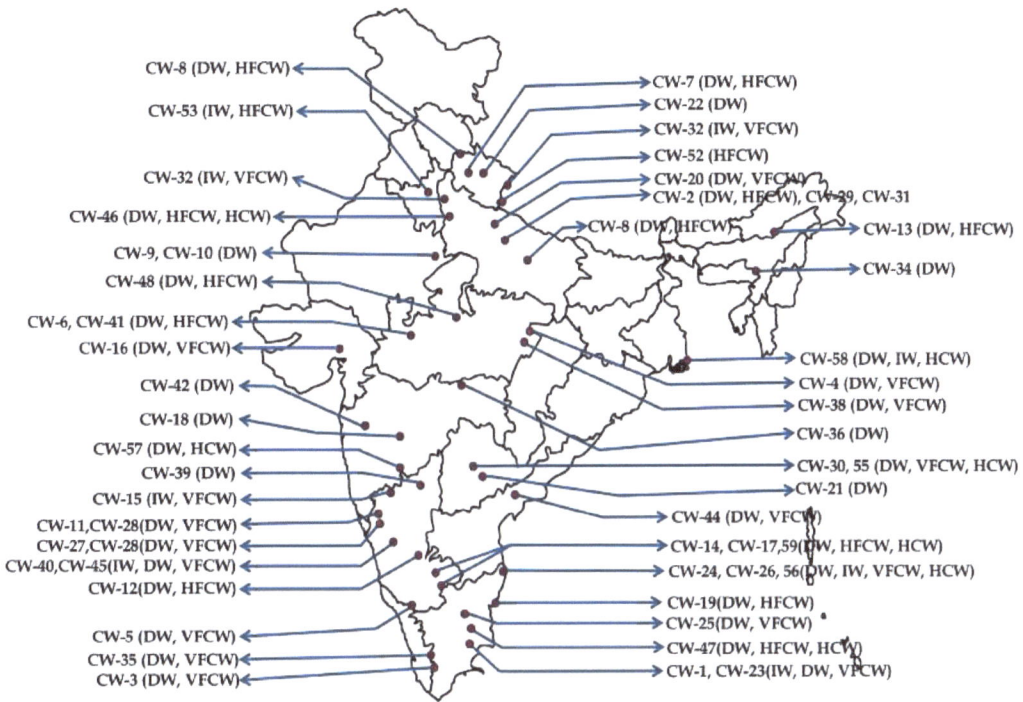

Fig. (1). The site distribution of constructed wetlands (CWs) in India. (Where: Constructed wetland number-1 (CW-1)), HFCW: Horizontal flow constructed wetland, VFCW: Vertical flow constructed wetland, HCW: Hybrid constructed wetland, DW: Domestic wastewater, IW: Industrial wastewater [3].

Plastics are generally synthetic or semi-synthetic materials manufactured by using polymers as main components. The word plastics was coined by Leo Hendrick Baekeland as he invented the fully synthetic plastic that is Bakelite. He is even termed as 'Father of the Plastic Industry' [4]. The word *plastic* derives from the Greek word (*plastikos*) meaning capable of being moldable and is made of synthetic organic polymers, which are usually produced through the polymerization of monomers derived from oil, gas, or coal [5]. The commercial production of plastic started in the 1940s and 1950s [6]. Plastics have brought great convenience to our daily lives but not without problems. The inappropriate disposal of wasted plastics has caused serious environmental problems. Over the past decade, microplastic debris are found in both marine and freshwater systems and it became an emerging issue [7]. All types of plastics are responsible for the degradation of the environment and have varied implications on different ecosystems; however, recently microplastics (MPs) have gained much interest in the scientific community. When plastic particles reach up to a size of <5mm, they are categorized as Microplastics [8]. 0.33 mm was defined as a lower limit for the size of microplastics based on the common mesh size of the Neuston nets which were used for the sampling of sea surface microplastics [9]. Our chapter deals with microplastics pollution in man-made water bodies along with its distribution in them.

TYPES OF WATER BODIES

A water body is any significant accumulation of water, generally on a planet's surface. Most are naturally-occurring geographical features, but some are artificial. There are types that can be natural as well as artificial. For example, most reservoirs are created by engineering dams, but some natural lakes are used as reservoirs. Similarly, most harbors are naturally occurring bays, but some harbors have been created through construction. Bodies of water that are navigable are known as waterways. Some bodies of water collect and move water, such as rivers and streams, and others primarily hold water, such as lakes and oceans.

Water bodies constitute areas of water – both salt and fresh, large and small – which are distinct from one another in various ways. From ponds to the Pacific, bodies of water rank among the most significant natural resources on the planet. The different types of water bodies are Inland wetlands and coastal wetlands. These are distinguished into Inland Natural & Inland man-made water bodies; Coastal Natural and Coastal Man-made water bodies. These are discussed as below:

Inland Water Bodies

These are classified in to inland natural and inland man-made wetlands.

Inland Natural Wetlands

The different types are discussed under:

Lakes

Larger bodies of slow moving or standing water occupying basins of appreciable size [10].

Ox-bow lakes/ Cut off Meanders

A meandering stream may erode the outside shores of its broad bends, and in time, the loops may become cut off, leaving basins. The shallow crescent-shaped lakes are called oxbow lakes. These are horseshoe shape lakes because the meandering stream continuously erodes outside the shores of its own broad bends and with the passage of time, the loops may become cut-off leaving basins [11].

Waterlogged Areas

An area in which water stands near, at, or above the land surface naturally by rainwater, termed as waterlogged area [12].

Playas

It is a temporary lake in deserts and often dries to form a salt pan. Playa basins have been divided into (1) the playas, which mark the fiat floor of the playa basin, (2) the annulus, which is considered the break in slope at the playa margin, and (3) the interplaya, which marks the region between the playas [13].

Swamps/Marshes

Along the major rivers, especially in plains, water accumulates leading to the formation of swamps and marshes. Marshes are frequently or continually inundated wetlands characterized by emergent herbaceous vegetation adapted to saturated soil conditions and swamps are wetlands dominated by trees or shrubs [14].

Rivers/ Streams

Rivers are linear water features of the landscape. Many stretches of the rivers in Indo-Gangetic Plains and Peninsular India are declared important National and International wetlands [15].

Inland Man-made Wetlands

Following types are discussed under:

Reservoirs

It is actually a pond or a lake built for the storage of water, usually by the construction of a dam across a river. There are different types of reservoirs that can hold gases, petroleum, diesel and water but the tank reservoirs for water are termed as cisterns.

There are different reasons for the formation of these water bodies. Some of these are discussed as:

A. To store water to use for irrigation, or watering crops.

B. People build reservoirs because the amount of water in a river varies over time. During very rainy times or when mountain snow is melting, the water in a river rises and sometimes overflows its banks. People build reservoirs in order to limit the amount of water which helps in controlling floods.

C. During droughts or extended dry periods, the water level in a river may be very low so water is released from the reservoir in the fields so farmers can have help in agriculture.

D. Reservoirs serve other purposes. They are used for boating, fishing, and other forms of recreation. Some of the dams that create reservoirs are used to generate electricity.

There are different types of reservoirs depending upon use like Valley-Dammed Reservoirs, Bankside reservoirs and Service Reservoirs.

Valley-Dammed Reservoirs

These dams are created in valleys between mountains. The mountain sides are used as the walls of the reservoir to hold water. A **dam**, or artificial wall in the reservoir, is built at the narrowest point to hold in the water.

Bank-side Reservoirs

These reservoirs are made by diverting water from local rivers or streams to an existing reservoir.

Service Reservoirs

The service reservoirs are entirely manmade and do not rely on damming a river or lake. These reservoirs are termed cisterns, and hold clean water. Cisterns can be dug in underground caverns or elevated high above the ground in a water tower [16].

Tanks / Ponds

Artificial ponds, pools or lakes are formed by building a mud wall across the valley of a small stream to retain monsoon. These may be very large or small water bodies specially built to retain rain water. The arid areas where rainfall is very low are very helpful in rain water harvesting.

Waterlogged

Man-made activities like canals cause waterlogging in adjacent areas due to seepage especially when canals are unlined. Paddy fields can be depicted as agricultural-managed temporary wetland ecosystems. These are actually man-made water bodies developed by modifying naturally existing waterlogged wetland areas. In addition to ensuring food security, other benefits of paddy fields include- flood control, water conservation and aquifer recharge [17].

Abandoned quarries

A quarry wetland is a water body that is formed after a quarry has been dug through a mining operation. A quarry is an area from which rocks such as marble, limestone, and granite are extracted for industrial use. The desired resources are taken and then, quarries are frequently abandoned. The resulting gaping holes can fill with water and form quarry lakes. Those are jointly termed as abandoned quarry [18].

Ash Pond/ Cooling Pond

An ash pond or cooling pond is a specially maintained structure used in fossil fuel power stations for the disposal of two types of coal combustion products: bottom ash and fly ash. The pond is used as a landfill to prevent the release of ash into the atmosphere and therefore helps in the reduction of pollution [19].

Coastal Wetlands

Classified into coastal natural and coastal man-made wetlands.

Coastal Natural Wetlands

Following are the types discussed below:

Estuary

An estuary is an area where a freshwater river or stream meets the ocean. When freshwater and seawater combine, the water becomes brackish, or slightly salty. It is actually the transition zone between the sea and freshwater [20].

Lagoons

Coastal bodies of water, partly separated from the sea by barrier beaches or bass of marine origin [21].

Creek

The creek is actually a tributary to the main river from which it occurs. These creeks develop with minor irregularities sooner or later causing the water to be deflected into definite channels [22].

Backwater / Baykayal

A backwater is part of a river in which there is little or no current. It can refer to a branch of a main river, which lies alongside it and then rejoins it, or to a body of water in the main river, backed up by the tide or by an obstruction such as a dam [23].

Bay

A bay is a body of water partly surrounded by land. A bay is usually smaller and less enclosed than a gulf. The mouth of the bay, which meets the ocean or lake, is typically wider than that of a gulf [24].

Mudflats/Tidal

Most un-vegetated areas that are alternately exposed and inundated by the falling and rising of the tide [25].

Sand beach

The beach is an unvegetated part of the shoreline formed of loose material, usually sand that extends from the upper berm [26].

Coral Reefs

Consolidated living colonies of microscopic organisms found in warm tropical waters [27].

Rocky Coast

The coast is the land along a sea. The boundary of a coast, where land meets water, is called the coastline. The Coastlines of granite, a hard rock, stay pretty stable for centuries. Such a type of rocky coastline is termed Rocky coast [28].

Mangroves

The mangrove swamp is an association of halophytic trees, shrubs, and other plants growing in brackish to saline tidal waters of tropical and sub-tropical coastlines. Mangrove forests provide food and shelter for diverse groups of fish and shellfish. These wetlands are often found in estuaries, where freshwater meets salt water and are infamous for their impenetrable maze of woody vegetation. By reducing the height and energy of wind and swell waves passing through them, mangroves perform as a cordon against the destructive force of nature like a tsunami and coastal erosion and reduces their ability to erode sediments and to cause damage to structure such as dikes and sea walls. Mangrove swamps are also natural sewage treatment plants [29].

Salt Marsh

Natural or semi-natural halophytic grassland and dwarf brushwood on the alluvial sediments bordering saline water bodies whose water level fluctuates either tidally or non-tidally. Saltwater and freshwater tidal marshes serve many important functions: They buffer stormy seas, slow shoreline erosion, offer shelter and nesting sites for migratory water birds, and absorb excess nutrients that would lower oxygen levels in the sea and harm wildlife [30].

Other Vegetation

The rest of the vegetation which does not find a place in any of the above-discussed types of wetlands, all the miscellaneous wetlands are kept with the other vegetation.

Coastal Man-made Water Bodies

Following types are discussed below:

Salt Pans

An un-drained usually small and shallow rectangular, man-made depression or hollow in which saline water accumulates and evaporates leaving a salt deposit. Recently with recent reports, these types are slowly losing their term of being wetlands [31].

Aquaculture Ponds

Aquaculture is defined as the breeding and rearing of fresh-water or marine fish in captivity, fish farming or ranching. The water bodies used for the above are called aquaculture ponds [32].

MICROPLASTICS OVERVIEW

Plastics are the most useful materials invented by man, and have brought great convenience in our day-to-day life. On one side, these are useful materials but on other hand have a number of problems. Plastic pollutants are variably classified according to the size, origin, shape, and composition. The plastics which are degraded to a size smaller than 5 mm are generally referred to as microplastic particles [33]. This is the biggest problem which is a gleaming topic among scholars around the world. Inappropriate disposal of wasted plastics has caused serious environmental issues. Microplastics are of global concern because they are present in most of the products which are used daily like personal care products: toothpaste, face scrubbers, lotions, soaps, *etc*. Microplastics are termed a wicked problem, *i.e.*, there is considerable complexity involved when one tries to understand the impact of these synthetic materials [7]. A higher level of microplastics is always generated as a result of the degradation of plastics. According to the literature, the amount of plastic entering water bodies is very high.

Plastic (from the Greek "plastikos", meaning mouldable) is made of synthetic organic polymers, which are usually produced through the polymerization of monomers derived from oil, gas, or coal [5]. The commercial production of plastic started in 1940s and 1950s [6]. Since then, plastic has extensively influenced human life, since it is used for a variety of purposes [34, 35] due to its outstanding qualities like light-weight, durability, versatility, and low production cost [5, 36]. However, there are many downsides to the present "plastic age", which include the extremely long half-life of plastics, their extensive use, and inefficient management of waste which cause an appalling build-up of these materials in the environment [37]. As plastics have gained importance in human life, concerns have been raised regarding the possible harmful impact on the environment and organisms which inhabit it, including humans.

All types of plastics are responsible for the degradation of the environment and have varied implications on different ecosystems; however, recently microplastics (MPs) have gained much interest of the scientific community. When plastic particles reach up to the size of <5mm, they are categorized as Microplastics. 0.33 mm was defined as a lower limit for the size of microplastics based on the common mesh size of the Neuston nets which were used for the sampling of sea surface microplastic [9]. However, this lower limit is not applied in recent microplastic studies, and particles as small as 1 μm in diameter are usually described as microplastics [5]. Additional classifications besides microplastics and macroplastics are also in vogue in literature. Recently, microplastics were discerned again into large microplastics (1-5 mm) and small microplastics (20μ - 1 mm) and nanoplastics (1-1000nm) in several studies [38]. The term mesoplastics is used to describe plastic particles in the size range of 5 to maximum 25 mm [38, 39]. Microplastics are classified as either primary or secondary. Primary MPs are designed and manufactured to be "micro-sized", whereas, secondary MPs are introduced through the process of fragmentation and degradation of large plastic materials. Secondary plastics are more liable to fragmentation by intense weathering and exposure to higher UV levels. MPs come from a variety of sources, including from large plastic debris that degrades into smaller and smaller fragments [40]. In addition, microbeads, a type of microplastic, are very tiny pieces of manufactured polyethylene plastic that are added as exfoliants to health and beauty products, such as in some cleansers and toothpaste. These tiny particles easily pass through water filtration systems and end up in the ocean and lakes, posing a potential threat to aquatic life.

MPs are also found in a wide range of shapes, *e.g.* spheres (beads), fibre, and film. These differences in the shapes and densities of MPs cause them to disperse differently in different regions or compartments of the aquatic environment *i.e.* water surface, water column and sediment and as such their availability to organisms at different trophic levels and/or ability to occupy different habitats is also influenced [41, 42, 43]. Microplastics are found in almost all marine and freshwater environments and have been detected in protected and remote areas making their potential detrimental effects a global problem. Our water bodies are getting contaminated with microplastics by different sources like household wastes, sewage wastes, and even from industrial effluents [38, 44]. The presence of microplastics as a new type of emerging contaminant has become a great concern for public and government authorities.

SOURCES OF MICROPLASTICS INTO MAN-MADE WATER BODIES

Plastic wastes are of high environmental concern due to their universal usage and distribution which is affecting not only terrestrial habitats but aquatic habitats too. There is always increasing attention to knowing how they find their way to water bodies and accumulate in the different water bodies. Freshwater microplastic pollution is a topic of concern globally, but its distribution and sources in reservoirs are poorly documented. In this sub-heading we will try to accommodate the sources of mps in man-made reservoirs or man-made water bodies. Plastics after entering water bodies eventually end up as persistent pollutants. They take a long time to degrade in nature, although plastics can still be fragmented into smaller pieces by a range of environmental factors. Compared to larger plastics, MPs may also be subject to different rates of degradation as they will be transported and distributed to various environment compartments at quicker rates than macroplastics. It is very much applicable in the oceans where changes in temperature, UV radiation, oxygen content and physical activity, *viz* wave action and ocean circulation can cause the fragmentation of plastics into tiny pieces [54]. The mechanical, chemical and biological degradation of plastics turns them into microplastics that are less than 5 mm in size and these are more harmful than large plastic debris [45, 46].

In other words, we can say, the degradation of large plastic debris is generally classified according to the medium which helps in causing it that are discussed below:

a) Biodegradation – Action of living organisms usually microbes on the plastics.

b) Photo-degradation – Action of light (usually sunlight in outdoor exposure).

c) Thermo oxidative degradation – Slow oxidative breakdown of larger plastic debris at moderate temperatures.

d) Thermal degradation – Action of high temperatures.

e) Hydrolysis – Reaction with water [47].

Plastics find their way to water bodies through different sources like industrial effluents, waste water treatment plant, agricultural wastes, domestic wastes, tyres, synthetic textiles, marine coatings, road markings, personal care products, plastic pellets, city dust, packaging materials and wrappers, consumer products *etc*.

Freshwaters represent the most complex system regarding microplastic transport and retention, as they receive microplastics from the terrestrial environment. Rivers play a very important role in transferring microplastics from terrestrial medium to water bodies (Fig. **2**) as larger plastic debris found their way directly in them from different anthropogenic activities as inadequate waste disposal, either through littering or loss from landfill and transported from land *via* wind or surface runoff. On the other hand, agricultural drainage along with runoff from farmland also adds agricultural plastics or sewage-sludge derived fibers and microbeads into the streams. Likewise storm drainage and urban runoff are often unfiltered and untreated, and can contain microplastics from degraded road paint and wear from vehicles [35, 48]. These fresh waterbodies act as a means of microplastic transporter through the breakup of larger items into smaller one as well as act as sinks which retain microplastics in their sediments. Instead of these, during very high flow circumstances, combined sewage overflows are designed to release untreated sewage into surrounding rivers to reduce the pressure on drainage systems, releasing both micro- and macroplastic waste. Many of the studies indicate that the hubs of microplastics may occur in close proximity to urban areas, the majority of microplastics are likely to enter waterbodies as a result of drainage systems [49].

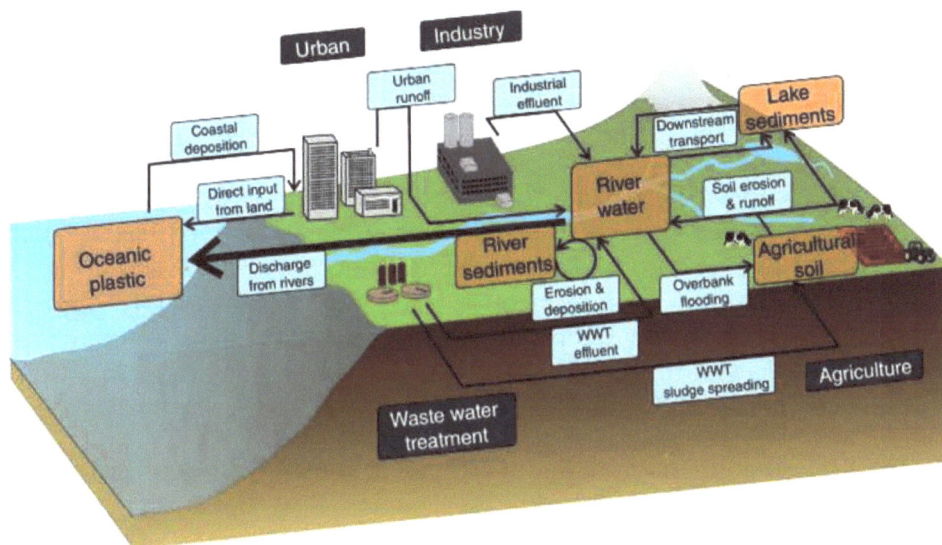

Fig. (2). Transportation of Microplastics in the environment.

The wastewater treatment plants are the main sources of microplastics to the water bodies [50]. Regardless of the efficiency of wastewater treatment plants in removing microplastics, direct effluent of wastes must contain microplastics that get their way into the water as these are very small and light in weight [51]. The fragments of plastics that are most likely the result of degradation or weathering of larger plastic products, transported over large distances by water masses or input of effluents from the wastewater treatment plants. Industrial activities are the other important matter of concern, which are a great source of plastic pollution in water bodies [52]. For instance, in Kerala, the majority of its industries are located in its coastal districts and riverbanks. The backwaters and major rivers of the region act as a channel for the discharge of dissolved and particulate matter from these industries to artificial water bodies [53].

Fibres are mostly originated from discarded rope material because of heavy marine traffic and fishing activities [47]. The Red hill lakes is one of the main drinking water suppliers to the residents of Chennai city. It is also known as Puzhal Aeri/Lake. Other than these, the two more reservoirs are there from which water is supplied for usage by the people of Chennai, those are Chembarambakkam Lake and Porur Lake. They are contaminated by different types of microplastics which are fibers, fragments, films and pellets. The main sources of these microplastics are mainly the fragmentation of plastic products due to weathering process and also fishing nets that are also vital contributors of microplastics in water and sediments [54]. Another possible source of microplastics in water bodies is dry deposition through the medium of wind. The microplastic pollution emitted from vehicle exhausts, and tyres also find their way from the terrestrial atmosphere to the aquatic habitat [55].

The Feilaixia Reservoir (Guangdong Province, 17 China) plays an important role in people's daily lives in Guangdong Province. An investigation has been carried out of the surface water of the said reservoir to find microplastics pollution. The source of microplastics in the surface water can be determined based on their morphological features. For instance, the fibrous microplastics mainly come from clothes, textiles, fishing nets and ropes, the microplastic films mainly come from disposable plastic bags, the microplastic fragments are mainly derived from plastic products made with thicker materials, and the foams are mainly derived from expanded polystyrene products [56, 57].

Similarly, work is done to know the microplastic contamination of Danjiangkou Reservoir which is the second largest reservoir in China and is divided into the Han Reservoir and Dan Reservoir. Microplastic abundances and morphological characteristics of the reservoir were investigated in that work. The fragments, micro-beads, and pellets were reported from the reservoir. The sources will be the same like clothes, textiles, fishing nets, agricultural wastes, and larger plastic debris which are degraded into smaller fragments [58].

DISTRIBUTION OF MICROPLASTICS ON MANMADE WATER BODIES ECOSYSTEM

The distribution of microplastics in aquatic environments is affected by a variety of factors like land use type, population density, hydrodynamic properties of water bodies such as tides and strong currents and sediment properties [42]. In spite of above discussed external factors, the physicochemical properties of microplastics may also affect their distribution and fate in the water bodies. The type of weathering and fragmentation of plastics also impact the distribution of microplastics in the aquatic environment. As already discussed, there are different types of water bodies, such as *Reservoirs* that are artificial water bodies created by damming a river for the purpose of flood control, electricity generation, irrigation, water supply and shipping. Mostly, the hydrodynamic properties of reservoirs are similar to that of a lake, since there is also decreased flow rate which makes these reservoirs a favorable destination for the accumulation of plastic debris. So the different types of microplastics take up different places in the water body. This depends upon the density, surface area, and size of plastics. The smaller microplastics are more likely to suspend in the upper layer of water column, although they sometimes sink to the bottom as a result of reactions like the formation of biofilm [59, 60]. Other than the size of the microplastics, the high-density microplastics may settle faster in comparison to low density microplastics since high-density microplastics usually have a smaller specific surface area [61]. Other than this, the spatial and temporal distribution of microplastics in water bodies is also being studied but there is very less information about it. Prominent work has been done on the Antua river of Portugal which aims to provide new insights into microplastics' abundance and distribution in the said river. The different stations and times of year were selected to know about the abundance and distribution of microplastics in the river. It has been found that the abundance of MPs exhibits different results in different seasons. Microparticles were found in all water samples with abundances ranging from 5 to 51.7 mg m^{-3} or 58–1265 items m^{-3}. By comparing all water samples, the highest MPs abundance was found during October and the lowest abundance was found during March. The spatial distribution of MPs in water decreased from upstream to downstream areas of the river in October and it is not clear in March. The different physico-chemical parameters along with hydrodynamics parameters and seasonal conditions of the water body may be the cause of such a distribution of microplastics in the said river [62]. Other factors like the vicinity of a waterbody like the industrial area, residential area whether urban or rural area or secluded area, *etc.* also have a great influence over the distribution and abundance of microplastics in them.

On the other hand, Lakes (natural or manmade) are shallow areas fed and drained by groundwater, rainfall or rivers in a catchment basin. Plastic waste generated within the lake catchment has the chance to be transported to the lake and accumulate there. An important work related to the distribution of microplastics has been carried out in different lakes and three great reservoirs namely (The three George Reservoirs of China). Microplastic pollution has been detected in Taihu Lake, urban lakes in Wuhan, lakes in Siling Co Basin, Qinghai Lake, and several lakes in Middle-Lower Yangtze River Basin. According to the spatial distribution of microplastics, a high abundance of microplastics in a plankton net sample was detected in the southeast area of the lake and high abundances of microplastics in the surface water and sediment samples were found in the northwest area of the lake. It was found that the microplastic distribution and abundance in the water body are negatively correlated with the distance from the city center. Even the lake currents and tourism also affect the distribution of microplastics due to which a higher concentration of microplastics is detected in the central areas of the lake [63].

Regarding reservoirs, Three Gorges Reservoir (TGR) is the largest reservoir in China. Three studies on microplastic pollution in Three Gorges Reservoir (TGR) have been reported. Microplastic abundance in surface water varied from 3407.7 × 103 to 13,617.5 × 103 particles/km^2 in the Yangtze River mainstream and 192.5 × 103 to 11,889.7 × 103 particles/km^2 in the tributary estuaries. The microplastic abundance showed an increasing tendency as the proximity to the dam increased.

Higher microplastic abundance levels were detected in the wet season as compared to those in the dry season due to the reason of different hydrodynamic conditions [63].

The mean microplastic abundance in the intermediate water is significantly higher than that of the surface water and bottom water. A report on the Danjiangkou Reservoir has shown that the presence of low-density particles decreased from the water surface to the subsurface up to the sediment, while the presence of high-density particles had the opposite result. In this freshwater reservoir, it has been reported that microplastics get accumulated in the middle-layer water. The surface currents in the Danjiangkou Reservoir have a higher velocity than the subsurface currents which is also one of the possible reasons for uneven distribution of microplastics that is, a high concentration in the intermediate zone of the reservoir and low concentration at the surface and sedimental area [58]. The horizontal distribution is another type of microplastic distribution. In one constructed wetland, it has been reported that with increasing distance from the shore, the microplastics concentration gets sparse. Microplastics that are transported may relate to the rate of biofilm attachment needed to sink the microplastics. More microplastics were transported further away from the inlet due to an increased number of larger aperture gaps between the substrate with coarse gravel which allowed microplastics to pass through [64].

In some studies, conventional statistical methods such as Pearson's correlation and log-linear correlations have been enforced to calculate the correlations between microplastic abundance and related factors [65, 66].

EFFECTS OF MICROPLASTICS ON MAN-MADE WATER BODIES ECOSYSTEM

Aquatic contamination by microplastics is now considered an emerging threat to biodiversity residing in the water bodies as well as the ecosystem functioning there because of the reason that this pollution affects the biology of different organisms as well as pollutes their habitat. Plastic pollution has been reported to negatively influence human health by causing toxicity through oxidative stress, inflammation and increased uptake or translocation [67]. The reported effects include abrasions and ulcers, blockages of the digestive tract, increased mortality, decreased fecundity, inflammatory response, alterations of metabolism, feeding modifications, reproductive disruption, changes in behavior, decreased energy reserves, the release of toxic monomers and/or additives, and vectors for other contaminants. Several studies have shown that plastics may be responsible for hormonal and metabolic disturbances, neurotoxicity and even increased cancer risk in humans. These negative effects always require higher concentration of microplastics in living organisms. The terrestrial ecosystems, especially agricultural land, have been identified as a major sink of microplastics, since from here, they become the source of microplastic pollution for the aquatic system by the ground water seepage to nearby water bodies to such areas. When the plastics enter into the reservoirs through various mediums, some young fish are attracted towards them because of their different colors and organisms assume it as food, and they end up consuming plastics as their natural food which eventually starves them or chokes them before they can reach their reproductive age [68]. Various studies have shown the presence of microplastics in the bodies of many aquatic organisms like fish, frogs, birds, and even in mammals (Fig. **3**). These actually pose a great threat to human health as a lot of human populations depend upon the big reservoirs for water supply [69].

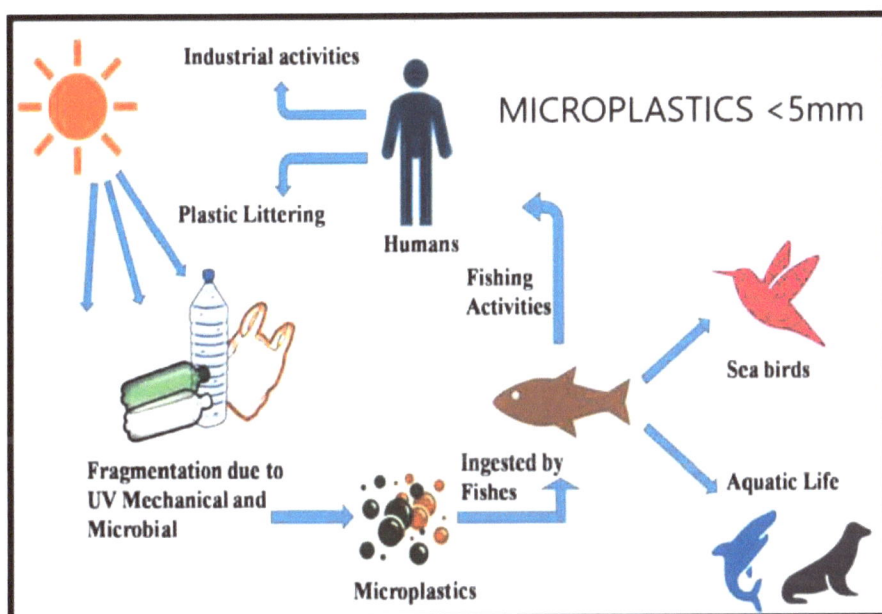

Fig. (3). Effects of microplastics on different organisms [70].

Many aquatic organisms including plankton, bivalves, fish, and mammals ingest the microplastics due to their small size, but they are unable to digest them. They even cause health consequences that are uncertain to the organisms; such deleterious effects are blockage, tissue damage and a false sense of fullness that limits nutrient uptake. This leads to a buildup of microplastics in the digestive tracts of the animals which hinders the microorganisms from taking in more food, and it diminishes the organisms feeding stimulus which can lead to starvation and ultimately death of the organism [71, 72]. Microplastics damage the lungs and other internal organs through accumulation in the respiratory and digestive tracts and some fine plastic particles produce toxic effects such as lower energy acquisition, oxidative stress, cell damage, and inflammation [73, 74]. As a secondary effect, ingestion of MPs may facilitate the transfer of persistent organic pollutants to other organisms as a possible trophic transfer. Many studies suggested that microplastics play a more consequential and important ecological part to play in the aquatic food web. The small size of microplastics facilitates easy uptake by organisms compared to larger ones. Microplastic ingestion has been reported for many different species, such as mussels, lugworms, crabs, seabirds, and fish [75, 76, 77, 78]. One of the major concerns among the scientific community regarding the presence of microplastics in various ecosystems is their bioaccumulation, in particular, bio-magnification and trophic transference. Microplastics have been detected in mussels, which are cultured for human consumption [79]. However, the retention time of microplastics in an organism depends on the uptake route as well as the type of the organism and if the retention time is long, the ingested microplastics can be transferred to the next trophic level [78]. The capability of MPs towards bioaccumulation increases with decreasing size [37]. Microplastics not only act as a primary pollutant but also have adsorption properties, so these can adsorb different heavy metals over them and act as secondary pollutants. It has been found that the microplastic pollutants with their adsorption property for heavy metals have combined toxic effects which include endocrine disruption and reduced predatory behavior of aquatic carnivores. Cadmium, cobalt, chromium, copper, nickel and lead are a few highly studied heavy metals that show interactions with microplastics [80]. Microplastics can also provide a peculiar microbial habitat, therefore acting as a carrier for pathogenic microorganisms and parasites in the aquatic environment which is unfit for use [81, 82].

CONCLUSION

The microplastic abundance in the intermediate zone of water body is significantly higher than that of the surface water and bottom water. It is due the reason that the surface currents in any reservoir have a higher velocity than the subsurface currents. The factors responsible for the distribution of microplastics in water bodies are land use type, population density, hydrodynamic properties of water bodies such as tides and strong currents as well as sediment properties even the physicochemical properties of microplastics also affect their distribution and fate in water bodies. So proper systematic studies should be done for estimating the distribution and fate of microplastics in the reservoirs. Even the studies related to the combined effects of multiple factors on the spatial distribution and temporal variation of microplastics in water bodies are still inadequate, and thus more systematic quantitative studies are urgently needed for understanding the spatial and temporal distribution of microplastics in them. It will be necessary to explicate the behaviors of microplastics in reservoir environments in the future. Other than the distribution, plastic pollution is on the rise nowadays because people are using plastic endlessly in the view of that plastics are very economical and easily available. Their disintegration and degradation generate microplastics that eventually lead to microplastic pollution which is a serious concern in the faculty of environmentalists these days. As the effects of microplastics are well studied on the biotic (living organisms) as well as abiotic (ecological aspects) factors of water bodies, so it is important to devise different realistic policies to combat and curb this so called Microplastic pollution. It is also important for government to enforce and implement regulations that will check production, consumption, usage and eventual disposal of plastics, irrespective of their hazardous status. The 3Rs: Reduce, Reuse, and Recycle must be employed at all stages so as to prevent zero diversion to landfills and indiscriminate disposal to the waterbodies. In spite of this, efforts must be made to educate general people on the potential environmental and public health effect of pollution by plastic wastes. This will go a long way to reduce the pollution rate and preserve the quality of the environment. There is need for people to be aware of the chemical constituents of plastic products and their health effects. Educational curriculums at different levels must include ways of plastic pollution reduction and waste management systems as information resources.

ACKNOWLEDGEMENT

Declared none.

REFERENCES

[1] Smith SV, Renwick WH, Bartley JD, Buddemeier RW. Distribution and significance of small, artificial water bodies across the United States landscape. Sci Total Environ 2002; 299(1-3): 21-36.
[http://dx.doi.org/10.1016/S0048-9697(02)00222-X] [PMID: 12462572]

[2] Gopal B. Natural and constructed wetlands for wastewater treatement: potentials and problems. Water Sci Technol 1999; 40(3): 27-35.
[http://dx.doi.org/10.2166/wst.1999.0130]

[3] Farheen KS, Reyes NJ, Kim LH. Constructed wetlands in treating domestic and industrial wastewater in india: a review. Journal of Wetlands Research 2021; 23(3): 242-51.

[4] Edgar D, Edgar R. Fantastic recycled plastic: 30 clever creations to spark your imagination.Sterling Publishing Company, Inc. 2009.

[5] Imhof HK, Schmid J, Niessner R, Ivleva NP, Laforsch C. A novel, highly efficient method for the separation and quantification of plastic particles in sediments of aquatic environments. Limnol Oceanogr Methods 2012; 10(7): 524-37.
[http://dx.doi.org/10.4319/lom.2012.10.524]

[6] Hahladakis JN, Velis CA, Weber R, Iacovidou E, Purnell P. An overview of chemical additives present in plastics: Migration, release, fate and environmental impact during their use, disposal and recycling. J Hazard Mater 2018; 344: 179-99.
[http://dx.doi.org/10.1016/j.jhazmat.2017.10.014] [PMID: 29035713]

[7]   Shilla D. Status updates on plastics pollution in aquatic environment of tanzania: data availability, current challenges and future research needs. Tanzan J Sci 2019; 45(1): 101-13.

[8]   Yousuf A, Naseer M, Naqash N, Singh R. Isolation and identification of microplastic particles from agricultural soil and its detection by fluorescence microscope technique. Think India Journal 2019; 22(16): 3934-49.

[9]   Eriksen M, Lebreton LCM, Carson HS, *et al.* Plastic pollution in the world's oceans: more than 5 trillion plastic pieces weighing over 250,000 tons afloat at sea. PLoS One 2014; 9(12): e111913.
[http://dx.doi.org/10.1371/journal.pone.0111913] [PMID: 25494041]

[10]   Lake Definition, Types, Examples, & Facts Britannica. Available at:https://www.britannica.com/science/lake

[11]   How Oxbow Lakes Form - Geography Realm. Available at:https://www.geographyrealm.com/how-oxbow-lakes-form/

[12]   Waterlog | Definition Of Waterlog by Oxford Dictionary on Lexico.com also meaning of WATERLOG. Available at:https://www.lexico.com/definition/waterlog

[13]   Scanlon BR, Goldsmith RS. Field study of spatial variability in unsaturated flow beneath and adjacent to playas. Water Resour Res 1997; 33(10): 2239-52.
[http://dx.doi.org/10.1029/97WR01332]

[14]   Marsh | National Geographic Society Available at:https://www.nationalgeographic.org/encyclopedia/marsh/

[15]   River and Stream | Encyclopedia.com. Available at:https://www.encyclopedia.com/science/encyclopedias-almanacs-transcripts-and-maps/river-and-stream

[16]   Liu W, Zhang Q, Liu G. Seed banks of a river–reservoir wetland system and their implications for vegetation development. Aquat Bot 2009; 90(1): 7-12.
[http://dx.doi.org/10.1016/j.aquabot.2008.04.014]

[17]   Chowdary VM, Chandran RV, Neeti N, *et al.* Assessment of surface and sub-surface waterlogged areas in irrigation command areas of Bihar state using remote sensing and GIS. Agric Water Manage 2008; 95(7): 754-66.
[http://dx.doi.org/10.1016/j.agwat.2008.02.009]

[18]   Tan F, Jiao YY, Wang H, Liu Y, Tian H, Cheng Y. Reclamation and reuse of abandoned quarry: A case study of Ice World & Water Park in Changsha. Tunn Undergr Space Technol 2019; 85: 259-67.
[http://dx.doi.org/10.1016/j.tust.2018.12.009]

[19]   Patel GK, Pitroda J. Assessment of natural sand and pond ash in Indian context. Int J Eng Trends Technol 2013; 4(10): 4287-92.

[20]   Pritchard DW. Pritchard DW. What is an estuary: physical viewpoint. American Association for the Advancement of Science.

[21]   Lagoon | National Geographic Society. Available at:https://www.nationalgeographic.org/encyclopedia/lagoon/

[22]   Healy TR. Tidal creeks.Encyclopedia of earth sciences series. Springer Netherlands 2005; Vol. 14: pp. 949-50.
[http://dx.doi.org/10.1007/1-4020-3880-1_313]

[23]   Backwater | Definition of Backwater by Merriam-Webster. Available at:webster.com/dictionary/backwater

[24]   Bay | National Geographic Society. Available at:https://www.nationalgeographic.org/encyclopedia/bay/

[25]   Murray NJ, Phinn SR, DeWitt M, *et al.* The global distribution and trajectory of tidal flats. Nature 2019; 565(7738): 222-5.

[http://dx.doi.org/10.1038/s41586-018-0805-8] [PMID: 30568300]

[26] Sand Beach Wetlands Conservation Area - Naturally Lennox & Addington Available at:https://naturallyla.ca/explore/trails-and-nature-parks/sand-beach-wetlands-conservation-area/

[27] Coral Reefs | National Geographic Society. Available at:https://www.nationalgeographic.org/article/coral-reefs/

[28] Coast | National Geographic Society. Available at:https://www.nationalgeographic.org/encyclopedia/coast/

[29] US EPA, O. Mangrove Swamps. Available at:https://www.epa.gov/wetlands/mangrove-swamps

[30] Wang HW, Kuo PH, Dodd AE. Gate operation for habitat-oriented water management at Budai Salt Pan Wetland in Taiwan. Ecol Eng 2020; 148: 105761.
[http://dx.doi.org/10.1016/j.ecoleng.2020.105761]

[31] What Are Salt Pans Land: Impact on Mumbai real estate. Available at:https://www.proptiger.com/guide/post/all-you-need-to-know-about-salt-pan-land

[32] Pond Culture – Freshwater Aquaculture. Available at:https://freshwater-aquaculture.extension.org/pond-culture/

[33] Arthur WB. The nature of technology: What it is and how it evolves. Simon and Schuster 2009.

[34] Barnes DKA, Galgani F, Thompson RC, Barlaz M. Accumulation and fragmentation of plastic debris in global environments. Philos Trans R Soc Lond B Biol Sci 2009; 364(1526): 1985-98.
[http://dx.doi.org/10.1098/rstb.2008.0205] [PMID: 19528051]

[35] Horton AA, Walton A, Spurgeon DJ, Lahive E, Svendsen C. Microplastics in freshwater and terrestrial environments: Evaluating the current understanding to identify the knowledge gaps and future research priorities. Sci Total Environ 2017; 586: 127-41.
[http://dx.doi.org/10.1016/j.scitotenv.2017.01.190] [PMID: 28169032]

[36] Hammer J, Kraak MH, Parsons JR. Plastics in the marine environment: the dark side of a modern gift. Rev Environ Contam Toxicol 2012; 220: 1-44.
[PMID: 22610295]

[37] Wagner M, Scherer C, Alvarez-Muñoz D, *et al.* Microplastics in freshwater ecosystems: what we know and what we need to know. Environ Sci Eur 2014; 26(1): 12.
[http://dx.doi.org/10.1186/s12302-014-0012-7] [PMID: 28936382]

[38] Rhodes CJ. Plastic pollution and potential solutions. Sci Prog 2018; 101(3): 207-60.
[http://dx.doi.org/10.3184/003685018X15294876706211] [PMID: 30025551]

[39] Isobe A, Kubo K, Tamura Y, Kako S, Nakashima E, Fujii N. Selective transport of microplastics and mesoplastics by drifting in coastal waters. Mar Pollut Bull 2014; 89(1-2): 324-30.
[http://dx.doi.org/10.1016/j.marpolbul.2014.09.041] [PMID: 25287228]

[40] Bellasi A, Binda G, Pozzi A, Galafassi S, Volta P, Bettinetti R. Microplastic contamination in freshwater environments: A review, focusing on interactions with sediments and benthic organisms. Environments 2020; 7(4): 30.
[http://dx.doi.org/10.3390/environments7040030]

[41] Betts K. Why small plastic particles may pose a big problem in the oceans. Environ Sci Technol 2008; 42(24): 8995.
[http://dx.doi.org/10.1021/es802970v]

[42] Cole M, Lindeque P, Halsband C, Galloway TS. Microplastics as contaminants in the marine environment: A review. Mar Pollut Bull 2011; 62(12): 2588-97.
[http://dx.doi.org/10.1016/j.marpolbul.2011.09.025] [PMID: 22001295]

[43] Thompson RC. Microplastics in the marine environment: sources, consequences and solutions.InMarine anthropogenic litter. Cham: Springer 2015; pp. 185-200.

[44] Tanaka K, Takada H. Microplastic fragments and microbeads in digestive tracts of planktivorous fish from urban coastal waters. Sci Rep 2016; 6(1): 34351.
[http://dx.doi.org/10.1038/srep34351] [PMID: 27686984]

[45] Klein S, Dimzon IK, Eubeler J, Knepper TP. Analysis, occurrence, and degradation of microplastics in the aqueous environment.InFreshwater microplastics. Cham: Springer 2018; pp. 51-67.

[46] Singh B, Sharma N. Mechanistic implications of plastic degradation. Polym Degrad Stabil 2008; 93(3): 561-84.
[http://dx.doi.org/10.1016/j.polymdegradstab.2007.11.008]

[47] Andrady AL. Microplastics in the marine environment. Mar Pollut Bull 2011; 62(8): 1596-605.
[http://dx.doi.org/10.1016/j.marpolbul.2011.05.030]

[48] Boucher J, Friot D. Primary microplastics in the oceans.IUCN, Global Marine and Polar Programme. IUCN Publication 2017; p. 43.
[http://dx.doi.org/10.2305/IUCN.CH.2017.01.en]

[49] Horton AA, Dixon SJ. Microplastics: An introduction to environmental transport processes. WIREs Water 2018; 5(2): e1268.
[http://dx.doi.org/10.1002/wat2.1268]

[50] Yan M, Yang J, Sun H, Liu C, Wang L. Occurrence and distribution of microplastics in sediments of a man-made lake receiving reclaimed water. Sci Total Environ 2022; 813: 152430.
[http://dx.doi.org/10.1016/j.scitotenv.2021.152430] [PMID: 34952049]

[51] Murphy F, Ewins C, Carbonnier F, Quinn B. Wastewater treatment works (WwTW) as a source of microplastics in the aquatic environment. Environ Sci Technol 2016; 50(11): 5800-8.
[http://dx.doi.org/10.1021/acs.est.5b05416] [PMID: 27191224]

[52] Manzoor S, Kaur H, Singh R. Existence of microplastic as pollutant in harike wetland: an analysis of plastic composition and first report on ramsar wetland of india. Current World Environment 2021; 16(1).
[http://dx.doi.org/10.12944/CWE.16.1.12]

[53] Robin RS, Karthik R, Purvaja R, *et al.* Holistic assessment of microplastics in various coastal environmental matrices, southwest coast of India. Sci Total Environ 2020; 703: 134947.
[http://dx.doi.org/10.1016/j.scitotenv.2019.134947] [PMID: 31734498]

[54] Manzoor S, Naqash N, Rashid G, Singh R. Plastic material degradation and formation of microplastic in the environment: A review. Mater Today Proc 2021.

[55] Dris R, Gasperi J, Saad M, Mirande C, Tassin B. Synthetic fibers in atmospheric fallout: A source of microplastics in the environment? Mar Pollut Bull 2016; 104(1-2): 290-3.
[http://dx.doi.org/10.1016/j.marpolbul.2016.01.006] [PMID: 26787549]

[56] Tan X, Yu X, Cai L, Wang J, Peng J. Microplastics and associated PAHs in surface water from the Feilaixia Reservoir in the Beijiang River, China. Chemosphere 2019; 221: 834-40.
[http://dx.doi.org/10.1016/j.chemosphere.2019.01.022] [PMID: 30684781]

[57] Manzoor S. Kaur H, Singh R. Analysis Of Nylon 6 As Microplastic In Harike Wetland By Comparing Its IR Spectra With Virgin Nylon 6 And 6.6. Eur J Mol Clin Med 2020; 7(07): 2020.

[58] Lin L, Pan X, Zhang S, *et al.* Distribution and source of microplastics in China's second largest reservoir - Danjiangkou Reservoir. J Environ Sci (China) 2021; 102: 74-84.
[http://dx.doi.org/10.1016/j.jes.2020.09.018] [PMID: 33637267]

[59] Dai Z, Zhang H, Zhou Q, *et al.* Occurrence of microplastics in the water column and sediment in an inland sea affected by intensive anthropogenic activities. Environ Pollut 2018; 242(Pt B): 1557-65.
[http://dx.doi.org/10.1016/j.envpol.2018.07.131] [PMID: 30082155]

[60] Kooi M, Nes EH, Scheffer M, Koelmans AA. Ups and downs in the ocean: effects of biofouling on vertical transport of microplastics. Environ Sci Technol 2017; 51(14): 7963-71.

[http://dx.doi.org/10.1021/acs.est.6b04702] [PMID: 28613852]

[61]   Schwarz AE, Ligthart TN, Boukris E, van Harmelen T. Sources, transport, and accumulation of different types of plastic litter in aquatic environments: A review study. Mar Pollut Bull 2019; 143: 92-100.
[http://dx.doi.org/10.1016/j.marpolbul.2019.04.029] [PMID: 31789171]

[62]   Rodrigues MO, Abrantes N, Gonçalves FJM, Nogueira H, Marques JC, Gonçalves AMM. Spatial and temporal distribution of microplastics in water and sediments of a freshwater system (Antuã River, Portugal). Sci Total Environ 2018; 633: 1549-59.
[http://dx.doi.org/10.1016/j.scitotenv.2018.03.233] [PMID: 29758905]

[63]   Zhang K, Shi H, Peng J, *et al.* Microplastic pollution in China's inland water systems: A review of findings, methods, characteristics, effects, and management. Sci Total Environ 2018; 630: 1641-53.
[http://dx.doi.org/10.1016/j.scitotenv.2018.02.300] [PMID: 29554780]

[64]   Chen Y, Li T, Hu H, *et al.* Transport and fate of microplastics in constructed wetlands: A microcosm study. J Hazard Mater 2021; 415: 125615.
[http://dx.doi.org/10.1016/j.jhazmat.2021.125615] [PMID: 33725550]

[65]   Amrutha K, Warrier AK. The first report on the source-to-sink characterization of microplastic pollution from a riverine environment in tropical India. Sci Total Environ 2020; 739: 140377.
[http://dx.doi.org/10.1016/j.scitotenv.2020.140377] [PMID: 32758976]

[66]   Corcoran PL. Degradation of microplastics in the environment.InHandbook of Microplastics in the Environment. Cham: Springer 2022; pp. 531-42.

[67]   Singh R, Manzoor S, Naqash N. Microplastic hazard, management, remediation, and control strategies: a review. Int J Environ Technol Manag 2022; 1(1): 10049175.
[http://dx.doi.org/10.1504/IJETM.2022.10049175]

[68]   Triebskorn R, Braunbeck T, Grummt T, *et al.* Relevance of nano- and microplastics for freshwater ecosystems: A critical review. Trends Analyt Chem 2019; 110: 375-92.
[http://dx.doi.org/10.1016/j.trac.2018.11.023]

[69]   Vijayaraman S, Mondal P, Nandan A, Siddiqui NA. Presence of microplastic in water bodies and its impact on human health. In Advances in Air Pollution Profiling and Control. Singapore: Springer 2020; pp. 57-65.

[70]   Issac MN, Kandasubramanian B. Effect of microplastics in water and aquatic systems. Environ Sci Pollut Res Int 2021; 28(16): 19544-62.
[http://dx.doi.org/10.1007/s11356-021-13184-2] [PMID: 33655475]

[71]   Fossi MC, Coppola D, Baini M, *et al.* Large filter feeding marine organisms as indicators of microplastic in the pelagic environment: The case studies of the Mediterranean basking shark (Cetorhinus maximus) and fin whale (Balaenoptera physalus). Mar Environ Res 2014; 100: 17-24.
[http://dx.doi.org/10.1016/j.marenvres.2014.02.002] [PMID: 24612776]

[72]   Hu L, Chernick M, Hinton DE, Shi H. Microplastics in small waterbodies and tadpoles from Yangtze River Delta, China. Environ Sci Technol 2018; 52(15): 8885-93.
[http://dx.doi.org/10.1021/acs.est.8b02279] [PMID: 30035533]

[73]   Eerkes-Medrano D, Thompson RC, Aldridge DC. Microplastics in freshwater systems: A review of the emerging threats, identification of knowledge gaps and prioritisation of research needs. Water Res 2015; 75: 63-82.
[http://dx.doi.org/10.1016/j.watres.2015.02.012] [PMID: 25746963]

[74]   Galloway TS. Micro-and nano-plastics and human health.InMarine anthropogenic litter. Cham: Springer 2015; pp. 343-66.

[75]   Besseling E, Wegner A, Foekema EM, van den Heuvel-Greve MJ, Koelmans AA. Effects of microplastic on fitness and PCB bioaccumulation by the lugworm *Arenicola marina* (L.). Environ Sci Technol 2013; 47(1): 593-600.

[http://dx.doi.org/10.1021/es302763x] [PMID: 23181424]

[76]   Lusher AL, McHugh M, Thompson RC. Occurrence of microplastics in the gastrointestinal tract of pelagic and demersal fish from the English Channel. Mar Pollut Bull 2013; 67(1-2): 94-9.
[http://dx.doi.org/10.1016/j.marpolbul.2012.11.028] [PMID: 23273934]

[77]   von Moos N, Burkhardt-Holm P, Köhler A. Uptake and effects of microplastics on cells and tissue of the blue mussel Mytilus edulis L. after an experimental exposure. Environ Sci Technol 2012; 46(20): 11327-35.
[http://dx.doi.org/10.1021/es302332w] [PMID: 22963286]

[78]   Watts AJR, Lewis C, Goodhead RM, *et al.* Uptake and retention of microplastics by the shore crab Carcinus maenas. Environ Sci Technol 2014; 48(15): 8823-30.
[http://dx.doi.org/10.1021/es501090e] [PMID: 24972075]

[79]   Van Cauwenberghe L, Janssen CR. Microplastics in bivalves cultured for human consumption. Environ Pollut 2014; 193: 65-70.
[http://dx.doi.org/10.1016/j.envpol.2014.06.010] [PMID: 25005888]

[80]   Naqash N, Prakash S, Kapoor D, Singh R. Interaction of freshwater microplastics with biota and heavy metals: a review. Environ Chem Lett 2020; 18(6): 1813-24.
[http://dx.doi.org/10.1007/s10311-020-01044-3]

[81]   Jiang JQ. Occurrence of microplastics and its pollution in the environment: A review. Sustainable production and consumption 2018; 13: 16-23.
[http://dx.doi.org/10.1016/j.spc.2017.11.003]

[82]   McCormick A, Hoellein TJ, Mason SA, Schluep J, Kelly JJ. Microplastic is an abundant and distinct microbial habitat in an urban river. Environ Sci Technol 2014; 48(20): 11863-71.
[http://dx.doi.org/10.1021/es503610r] [PMID: 25230146]

SUBJECT INDEX

A

Acetylcholinesterase 50, 129
Acidobacteria 132
Acid(s) 7, 56, 138, 157, 162, 182
  humic 138
  hydrochloric 7
  nitric 7
  polylactic 56, 157, 162
Activities 21, 28, 51, 57, 130, 132, 133
  enzyme 57, 132
  immune enzyme 51
  isopod drilling 21
  metabolic 51
  microbial 28, 57, 130, 133
Adsorption 81, 84, 137, 214
  processes 81, 84, 137
  properties 81, 214
  technique 81
Aggregation 54, 108, 130, 156, 158
  hemocyte 158
Agriculture, sustainable 125
Air 2, 25, 46, 47, 72, 73, 97, 99, 102, 177, 179, 180, 182
  ecosystems 73
  microplastic deposition 25
  polluted 99
Analysis, thermogravimetric 187
Anthropogenic 29, 45, 99, 129, 137, 197, 198, 208
  activities 29, 45, 129, 137, 197, 198, 208
  colloids 99
Anti-aggregation agents 177
Arbuscular mycorrhizal fungi (AMF) 57, 58, 131, 142
Ash, fly 203
Atmospheric microplastics 27, 29, 30
Atomic force microscopy (AFM) 106, 114, 115
Attenuated total reflection (ATR) 9, 184

B

Bacillus 131, 161
  cereus 131
  megaterium 161
Bacteria 166, 178
  contaminated 178
  photosynthetic 166
Bacterial polyesters 160
Bioaccumulation processes 75
Biochar adsorption 71, 84
Biodegradable plastic 22, 156, 157
Biodegradation 27, 32, 53, 97, 100, 165, 166, 167, 208
  aerobic 166
  anaerobic 167
Biofilter biodegradation ingestion 79
Bioplastic(s) 157, 159, 161, 164, 165, 166, 167
  biodegradation of 164, 165, 166, 167
  degradation 165
  non-biodegradable 157
  plant-based 159, 161, 167
  production 167
  synthesis 159
  waste 167
Breast cancer 158

C

Calcium metabolisms 158
Carcinogens 136
Cellulose 7, 163, 182
  acetate (CA) 7, 182
  based bioplastics 163
Chemical(s) 57, 75, 78, 101, 136, 137, 187
  analysis of polymer 78
  contaminants 75
  hydrophobic 187
  pollutants 137
  toxic 57, 101, 136
Chitinase 7, 182

W

www.ingramcontent.com/pod-product-compliance
Lightning Source LLC
Chambersburg PA
CBHW050831220326
41598CB00006B/352